Sustainability Solutions

An Interdisciplinary and Transdisciplinary Nexus

Sustainability Solutions addresses a wide range of global challenges with applied solutions in the interdisciplinary and transdisciplinary sustainability nexus. Titles in the series present a strong sustainable development focus, and topics which connect environment, economic and social applications are encouraged. Topics could include environmental impacts, vulnerability, risk assessment and reduction, human-environmental systems (e.g. food, water, health) and innovative solutions to climate change. Volumes in the series present novel solutions to societal challenges using innovative application, for example demonstrating a solutions-based approach using machine learning, artificial intelligence, decision support systems, advance geospatial technology. Case studies and shorter report volumes which present knowledge transfer for policymaking and governance are also encouraged.

Keywords:

Interdisciplinary environmental solutions, transdisciplinary knowledge transfer, geospatial technologies, natural resource management, decision support system, artificial intelligence, deep learning, smart technology, environmental science, economic, social and spatial, human-environment interactions, sustainable development

Rakesh Kumar
Editor

Sustainable Finance

Strategies and Tools to Manage Climate Risk

Springer

Editor
Rakesh Kumar
Department of Management Studies
Graphic Era Deemed to be University
Dehradun, Uttarakhand, India

ISSN 3059-278X ISSN 3059-2798 (electronic)
Sustainability Solutions
ISBN 978-3-031-97961-3 ISBN 978-3-031-97962-0 (eBook)
https://doi.org/10.1007/978-3-031-97962-0

This Springer imprint is published by the registered company Springer Nature Switzerland AG
The registered company address is: Gewerbestrasse 11, 6330 Cham, Switzerland

Preface

The book *Sustainable Finance: Strategies and Tools to Manage Climate Risk* examines smart city development, sustainable finance legislation, and sustainable energy solutions. It talks about the advantages of green bonds and climate bonds as well as how sustainable growth depends on renewable energy sources. The book offers comprehensive information about sustainable finance investment through sustainable technology and evaluates current policies on sustainable finance, including green bonds, sustainable investment methods, and environmental, social, and governance (ESG) factors.

The book explores the history of industry, regulatory changes, and innovative financial instruments like carbon tax, and green bonds. It highlights emerging trends in sustainable development, such as technology adaptation, climate risk assessment, and social bond growth. It emphasizes the importance of sustainable finance in addressing global challenges like climate change and social inequality. Every country is focusing on renewable energy as fossil fuels are depleting and energy demand is increasing.

The book explores the link between Green Finance (GF) principles and Sustainable Development strategies, focusing on how funding for environmentally friendly projects advances overall developmental goals. It provides a comprehensive analysis of case studies, and empirical data to clarify opportunities, challenges, and strategies for utilizing GF to hasten the transition to Sustainable Development. Smart technology integration in digital currency transactions is transforming financial ecosystems, lowering costs, increasing speed, and improving transparency. Blockchain, AI, the Internet of Things, and decentralized finance are major innovations transforming traditional systems.

AI will enhance fraud detection, facilitate autonomous payments, and boost supply chain management. Future trends include decentralized finance, cross-chain solutions, NFTs, and central bank digital currencies. Proactive cybersecurity, regulatory, and technological advancement strategies are necessary to navigate the constantly shifting landscape of digital finance. AI can support sustainable investment policies in renewable energy, but challenges include data quality, costs, regulatory issues, ethical issues, and scaling AI-based predictive models.

This book explores the role of fintech in advancing financial inclusion, highlighting its potential to reduce barriers to financial services and increase access for marginalized individuals. Financial inclusion boosts economic growth, poverty alleviation, and social development. The book examines the implementation of fintech, government promotion, and impact evaluation methods. It also explores how fintech supports the Sustainable Development Goals and provides insights for researchers, educators, policymakers, and practitioners. Given that it provides real-time scenarios of sustainable finance investment for lowering climate risk, it is helpful for professionals, policymakers, scholars, postgraduates, and undergraduate students. The book offers insightful information on achieving Sustainable Development Goal 2030.

Dehradun, India Rakesh Kumar

Introduction

The book title *Sustainable Finance: Strategies and Tools to Manage Climate Risk*, explores approaches to reduce climate risk. Sustainable development goal (SDG-2030) is a flagship agenda of United Nations (UN). This book examines the opportunities and challenges of sustainable finance in climate action plans. The book delved into sustainable energy, smart city development, and sustainable policies at the international level. It includes e recent case studies problems and, its solutions. The renewal energy resources are limited in a way, and conservation of these resources is a big challenge. This book focuses on the dependency of renewable energy resources for sustainable development. This project focuses on society development through the integration of technology involvement in climate action plans through sustainable finance. This examines the benefits of green bonds and climate bonds in sustainable development goals.

This project opens up policies on sustainable finance national and international level, like green bonds, sustainable investment strategies, and ESG factors between and among the countries. It assess existing climate policies and the risks involved. This book focuses on how sustainable finance policies can be transformed from traditional energy to renewable energy sources. This book assesses existing policies on sustainable finance at the local and international levels aligned with SDG-2030.

This book provides in-depth knowledge about sustainable finance investment through sustainable technology, which reduces climate risk. There are many challenges and opportunities for sustainable finance for the climate action plan, including carbon tax policies, ESG policies, technological risk, and many more. This project provided comprehensive knowledge for academicians, researchers, industrialists, and other stakeholders involved in sustainable finance businesses, including investment.

The book explores existing policies of sustainable finance and its futuristic challanges. It focuses government incentive for financial investment, challenge for sustainable finance investment and its solution. Its focuses on sustainable industry, renewal energy and sustainable technology which aligned with SDG-2030. This book helpful for professional, policies maker, scholar, postgraduate and undergraduate students. The reader will get real-time situation of sustainable finance investment for reducing of climate risk.

Contents

About the Editor

Dr. Rakesh Kumar is an eminent educationist and a decorated army officer with 25 years of experience. He is a classic example of courage, commitment, and an intimate relationship. His areas of expertise are Finance, Business Law, Labor Law, Technology, and Sustainability. He is an expert in Structural Equation Modeling (SEM), ISM, and Fuzzy TOPSIS research Methodology. He is an interdisciplinary researcher. He is also involved in designing and developing the course and syllabus on various Management topics, including analytics and the management curriculum. He has published more than 73 research papers in leading databases. He is the editor of a reputed book, including Taylor & Francis, Nova Science, and Bentham. He has published 33 Indian Patents. He has been part of the board of Study, the board of examination, and the Evaluation department in a government organization.

Chapter 1
Introduction and Brief History of Sustainable Finance

Vasim Ahmad, Anu Sayal, Rakesh Kumar, Amar Johri, Mohammad Asif, and Moin Uddin

Abstract Chapter gives a good overview of the complex ways in which sustainable finance comes into existence, develops, and is currently shaped with its definitional parts and imminent prospects for the short run. The essence of sustainable finance comprises the inclusion of ESG in financial decisions to facilitate long-term economic growth within an effectively managed natural environment under the mechanisms of social equity and good governance. It develops the history of the industry by discussing the impact of changes in regulation and new innovative financial instruments, like green bonds or ESG funds. The chapter also presents case studies of some of the early adapters, featuring their success stories and lessons learned from such pioneer attempts at integrating sustainability into financial practice. It exposes some of these criticism points, discussing how responses to such criticisms and adaptations have been forthcoming in dealing with the former aspect especially with regard to ESG reporting standards and regulatory improvements and has identified emerging trends and innovation occurring within the field of sustainable finance, such as increased technology adaptation, climate risk assessment, and the growth of

V. Ahmad (✉)
Uttaranchal Institute of Management, Uttaranchal University, Dehradun, India
e-mail: drahmadvasim@gmail.com

A. Sayal
School of Accounting and Finance, Taylor's Business School, Taylor's University, Subang Jaya, Malaysia
e-mail: anu.sayal@taylors.edu.my

R. Kumar
Graphic Era (Deemed to be) University, Dehradun, India
e-mail: aigrakeshkumar@gmail.com

A. Johri · M. Asif · M. Uddin
College of Administrative and Financial Sciences, Saudi Electronic University, Riyadh, Saudi Arabia
e-mail: a.johri@seu.edu.sa

M. Asif
e-mail: masif@seu.edu.sa

M. Uddin
e-mail: m.uddin@seu.edu.sa

R. Kumar (ed.), *Sustainable Finance*, Sustainability Solutions,
https://doi.org/10.1007/978-3-031-97962-0_1

social bonds. It speculates on possible future developments and again emphasizes the fact that sustainable finance before the background of rising global challenges due to climate change and social inequality is an indispensable tool. It highlights again that sustainable finance is among those transformational forces through which the financial sector can really lead systemic transformation toward building economy.

Keywords Sustainable finance · ESG criteria · Corporate social responsibility (CSR) · Impact investing · Triple bottom line · Green bonds · Social bonds · Sustainability reporting · Climate risk assessment

1.1 Introduction

Sustainable finance is the integration of ESG parameters in financial decision-making, pro-growth but ambient and socially caring, with fit structures of governance in the long run. It encompasses a broad spectrum of financial operations, such as investing, lending, insurance, having sustainable development goals at the pinnacle. It gears financial markets toward sustainability imperatives as a contribution to a low-carbon, inclusive, and resilient economy transition. The global context could never require sustainable finance more than at the present. Climate change, social inequalities, and failures in governance create major risks to economic growth globally and well-being within society (Schoenmaker and Schramade 2019). The financial sector is integral to addressing these challenges as it has the ability to mobilize and allocate capital. Climate Change and Environmental Degradation: An increased frequency and severity of climate-related disasters are urgently calling for action on climate change (TCFD 2017). Sustainable finance would, in this case, facilitate investments in renewable energy, efficiency, and other green technologies mitigating climate change and fostering environmental sustainability. Projects addressing such social challenges and contributing to more inclusive growth and societal resilience are financed by sustainable finance. (a) Questions of governance and ethics: A number of corporate scandals and failures in recent governance forcefully brought home the reasons for vigorous governance frameworks (OECD 2020). ESG criteria drive transparency, accountability, and good business practices as key drivers of investor trust and market integrity (Eccles et al. 2012).

In particular, the idea of sustainable finance has been identified to have its roots in early practices and concepts that placed ethical considerations at the core of financial decision-making. For example, the Quaker's negative screening policy against investment in the slave trade during the eighteenth century gave grounding for modern socially responsible investing. Like the Methodist Church in the USA in the nineteenth century, it instituted guidelines that ensured avoidance of investment in those industries initially judged to be sinful domains: alcohol, gambling, and tobacco (Renneboog et al. 2008). At the beginning of the twentieth century, ethical investing started to come to the fore with the formation of mutual funds based on ethical parameters. The Pioneer Fund, founded in 1928, is probably one of the very first mutual funds that adopted an investment policy excluding companies involved in alcohol, tobacco, and gambling from its portfolio. This really marked the moment when non-financial variables started being taken into account in investment decisions, highlighting or reaffirming the questions of morals and ethics in finance. The development of sustainable finance has been punctuated by a number of important milestones that shaped its evolution and adoption (Sparkes 2002).

The Exxon Valdez oil spill in 1989 brought to the forefront environmental risk exposure emanating from corporate operations and compelled investors to factor in environmental sustainability as one of the key investment criteria. This event launched the process of elaboration of ESG investing frameworks by already directly integrating the above-mentioned three pillars into the appraisal of investment opportunities. PRI has established a complete framework into how ESG issues should be included in investment practices, further supported by a network of signatories who are committed to responsible investing (Kurtz 2008). By December 2020, the PRI had signed up over 3000 signatories whose assets under management exceeded a whopping $100 trillion (PRI 2020). Now it was widely accepted by all, late recognition of sustainable finance as a practical, much-mainstream approach to finance. Much of the journey of sustainable finance has been aptly guided and, in turn, catalyzed by a few individuals and organizations who have been loud in their advocacy in its favor. An individual in this case, much acclaimed for pioneering the field of impact investing, is Sir Ronald Cohen, who is considered by most to be the godfather of this field. Cohen co-founded Apax Partners in 1972 and immediately became a key influence in the formation of the G8 Social Impact Investment Taskforce with its mandate to catalyze the global growth of "impact investing" (Cohen 2014). He is been at the forefront of incorporating social and environmental impacts into investment strategies.

The other influential organization is the Global Reporting Initiative (GRI), established in 1997. It designed the overall framework of sustainability reporting that now allows organizations to report on their ESG performance transparently. It is now a standard way of reporting on sustainability performance across the world, enabling investment decisions based on transparent and material ESG information (GRI 2013). In addition, the Carbon Disclosure Project, which was founded in 2000, heightened corporate transparency regarding climate risks and opportunities (CDP 2020). Through the provision of a platform for companies to disclose greenhouse emissions and climate strategies, CDP raised investor awareness on climate risks and developed climate considerations as aspects of investment decisions.

1.1.1 Historical Timeline and Significant Events

Various are the events that mark the history of sustainable finance. The modern era in this field was truly in the 1960s and 1970s, running parallel to the growth of more socially oriented movements and a greater sensitivity to environmental degradation. The Interfaith Centre on Corporate Responsibility is formed in 1971, moving shareholder activism in the direction of ethical investing. Several pioneering ethical funds were launched in the 1980s, and other earlier examples date back to 1971, like the Pax World Fund, and another in 1984, the Domini Social Equity Fund, which shunned companies involved in activities that improvised its ethical standards (Sparkes 2002). The Exxon Valdez oil spill of 1989 was to act as an inflection point in this field, after which hiding behind investment considerations without factoring in environmental variables could not be afforded, and ESG frameworks began to be accepted more globally (Kurtz 2008).

In the 1990s, the drive toward sustainable finance gained even greater momentum through the establishment, in 1997, of the Global Reporting Initiative, which offered standardized guidance on sustainability reporting (GRI 2013). The turn of the millennium, in 2000, saw the founding of the Carbon Disclosure Project, an organization set up to manage the disclosure of corporations. Most notably, the groundwork for the formalizing mechanism of sustainable finance was laid in 2006 with the establishment of the United Nations Principles of Responsible Investment (CDP 2020). The PRI framework encouraged a redefinition of the investor framework related to the integration of ESG drivers into investment decisions. Regulatory and policy developments have further been at the forefront of driving progress in sustainable finance. It was only in 2018 that the European Union saw its Sustainable Finance Disclosure Regulation come into force, under which financial market participants are required to disclose how they would integrate sustainability risks and opportunities into their investment decisions (PRI 2020). The latter should provide a basis for higher transparency and comparability in sustainable investments and, as a consequence, underpin increased confidence for the investor.

Similarly, the TCFD—Task Force on Climate-related Financial Disclosures—established in 2015 under the Financial Stability Board, made some recommendations for companies to disclose climate-related financial risks (European Commission 2018). The framework of TCFD got significant support from some investors and regulators that climate factors should be combined into the process of finance decisions (TCFD 2017). According to TCFD, in the USA, ESG disclosures are now somewhat more emphasized by the Securities and Exchange Commission. In 2021, the SEC announced the commission of a Climate and ESG Task Force to create enforcement related to ESG disclosure requirements, therefore increasing the regulatory line between sustainable finance and markets (TCFD 2017). One of the most important ways in which sustainable finance has been produced is in the sector of financial instruments. One of the first key sustainable financial instruments was the green bond. The European Investment Bank launched them in 2007 (SEC 2021). These bonds are particularly ring-fenced and used in projects that would contribute to an environmentally positive impact or form part of renewable energy, energy efficiency, or sustainable infrastructure. The dramatic growth has also been seen in ESG funds (Climate Bonds Initiative 2020). Such funds consider environmental, social, and governance criteria in the investment approach and aim to agree on financial returns with a general positive impact on society and the environment. According to Morningstar, global sustainable fund assets hit $1.7 trillion at the end of 2020, a rise driven by increasing investor demand for ESG investment options. Another

Table 1.1 Historical timeline of sustainable finance

Year	Event/Development	Description
1971	Pax World fund	Establishment of one of the first socially responsible investment funds
1982	Calvert social investment fund	Launch of a mutual fund focusing on socially responsible investing
1989	Exxon Valdez oil spill	Catalyzed the integration of environmental considerations into investment decisions
1997	Global reporting initiative (GRI)	Provided standardized guidelines for sustainability reporting
2006	UN principles for responsible investment (PRI)	Framework encouraging ESG integration in investment practices

key innovation is that of social bonds, which raise finance for projects with positive social outcomes (Morningstar 2021). In 2017, the International Capital Market Association released the Social Bond Principles, guiding the issuance of social bonds and attempting to introduce transparency and accountability into the process (ICMA 2017). Another quickly rising instrument is sustainability-linked loans, in which loan conditions, such as interest rates, are directly linked to the borrower's performance with respect to pre-set targets on sustainability. In other words, SLLs create incentives for the borrower to improve their performance on sustainability scores while further aligning the financial and sustainability goals (Loan Market Association 2019) (Table 1.1).

There has been a historical timeline full of significant events, regulatory and policy changes, and the development of new instruments that characterized the evolution of sustainable finance. The developments, therefore, provided an accumulated impetus toward the integration of sustainability considerations into financial markets and thereby driving growth and maturation in sustainable finance.

1.1.2 Objectives of the Chapter

The chapter is intended to provide an overview of the subject of sustainable finance, ranging from its historical development and important concepts to future directions. More precisely, this involves the following objectives:

1. Origins of Sustainable Finance: Precursor practices and milestones that formed the sustainable finance field, focusing on influential figures and organizations behind its development.
2. Evolution of Sustainable Finance: Analyze the historical development and major occasions that have contributed to the increase in sustainable finance, succeeded by a shift in regulations.

3. Key Concepts and Principles: Describe the core concepts underpinning Sustainable Finance, such as ESG criteria.
4. Case Studies: Examples will be presented of early movers in sustainable finance, with attendant successes and lessons learned from experience.
5. Challenges and Criticisms: To discuss the common obstacles faced by this industry, criticisms, and controversies, and how responses and adaptations have been conducted for improving these.
6. Future Directions: In this task, identify emerging trends and innovations in sustainable finance, and speculate about future developments for the field.

1.2 Environmental, Social, and Governance (ESG) Criteria

ESG refers to a set of standards against which the sustainability of the operations of a company is evaluated, and which socially conscious investors use in screening potential investments. The criteria apply to uncovering risks and opportunities that bear on a company's long-term sustainability and ethical impact.

Environmental Criteria: These have to do with how a firm acts a steward of nature. Some of the factors include energy use, waste, pollution, and conservation of natural resources, treatment of animals among others. For example, firms can be assessed in terms of their carbon footprint, water usage, impact on land/sea/air, and the ecological effects of their activities and products (Sievänen et al. 2013).
Social Criteria: These assess how it deals with its relations regarding workers, suppliers, customers, and the communities where it operates. The social issues here are very broad; they include human rights, supply chain labor standards, work place health, safety, and community engagement (Friede et al. 2015).
Governance Criteria: Companies' leadership, executive compensation, auditing, internal controls, and rights of shareholders are the criteria falling under this category. This is what investors seek in any company: management quality, board integrity, and the condition of shareholder relations (Gond et al. 2012).

1.2.1 Corporate Social Responsibility (CSR)

It essentially holds the concept that, besides maximizing profits, corporators should run after some other prosocial objectives. So, this category will include a wide range of activities such as:

Philanthropy: Many businesses donate or contribute to charities and non-profits in the interest of societal well-being (Carroll 1991).
Ethical Labor Practices: Fair and ethical treatment of workers.
Volunteering: This would entail encouraging employees to volunteer in doing community service.

Table 1.2 Sustainable financial instruments

Instrument	Description	Key features
Green bonds	Bonds specifically earmarked to fund projects with environmental benefits	Renewable energy, energy efficiency
ESG funds	Investment funds that integrate ESG criteria into their investment processes	Diversified portfolio, long-term sustainability focus
Social bonds	Bonds that finance projects with positive social outcomes	Affordable housing, healthcare, education
Sustainability-linked loans (SLLs)	Loans with terms tied to the borrower's performance against predefined sustainability targets	Incentives for improving ESG performance

Environmental Efforts: This does lower companies' carbon footprints and other adverse environmental impacts.
CSR initiatives can bolster a company's brand image and reputation, attract and retain talent, and promote customer loyalty (Porter and Kramer 2006).

1.2.2 *Impact Investing*

This can be targeted at a diversity of outcomes, including but not limited to:

Environmental Conservation: Investments in affordable and clean energy, sustainable agriculture, and water conservation.
Social Impact: Invest in affordable housing, healthcare, and education.
Economic Development: Investments in the field of microfinance, small and medium-sized enterprises, and other activities that enhance economic development are also some promising areas for impact investing (Bugg-Levine and Emerson 2011). The investor seeks active placement of capital in organizations or projects that further his social and environmental gains. The next key feature is measurement; hence, the impact investor has to cause a measure based on the frameworks and metrics to assess and report results for social and environmental performance of the investments. (O'Donohoe et al. 2010) (Table 1.2).

1.2.3 *Triple Bottom Line (People, Planet, Profit)*

The triple bottom line (TBL) is a framework for sustainability used in viewing the context of a company's social, environmental, and economic impact. This concept urges businesses to consider their impacts on society and the environment, not just profits.

People (Social Equity): This dimension deals with the general impact of business on employees, customers, suppliers, and the community, ensuring fair treatment, human rights, and social justice (Elkington 1998).
Planet (Environmental Responsibility): This aspect amounts to taking care of the environment through sustainable practices and involves reducing the ecological footprint and natural resource use through proper energy use and waste production management (Slaper and Hall 2011).
Profit (Economic Prosperity): This dimension is concerned with the economic value created by the organization. It emphasizes that business has to be economic to be able to be sustainable but has to be achieved in a social and environmental sensitive way (Norman and MacDonald 2004).

1.3 Case Studies

1.3.1 The Pax World Fund

The Pax World Fund, for instance, was established in 1971 and is commonly cited as among the first fund based on socially responsible investment. Founded by Luther Tyson and Jack Corbett—both were United Methodist ministers—the fund was formed with a key objective to provide investors a vehicle in which to avoid supporting industries such as weapons manufacturing and tobacco. The Pax World Fund was groundbreaking in incorporating ethics into the investment process and proved that one could do so without giving up financial returns (Pax World 2020).

1.3.2 Calvert Investments

Another early pioneer in this area was Calvert Investments, established in 1976. In 1982, it introduced the Calvert Social Investment Fund, said to be the first mutual fund intended for socially responsible investment. Calvert Investments spent years refining the stringent screening methodologies that companies would be gauged against in their ESG practices. This approach helped not only to exclude companies involved in harmful activities but also to actively include those that made positive contributions to society and the environment (Sparkes 2002).

1.3.3 Storebrand

Storebrand is a Norwegian financial services company and has been one of the first followers of the principles of sustainable finance. Already in 1995, the Storebrand Environmental Value Fund was launched to target investments in companies with superior environmental performance. Today, Storebrand holds a position as one of the leaders in responsible investment practice (Storebrand 2021).

1.3.4 Success Stories and Lessons Learned

1. One of the most famous success stories in sustainable finance, Generation Investment Management integrates sustainability research into traditional financial analysis to identify long-term investment opportunities, emphasizing that ESG factors are an integral part of a holistic investment strategy.

Key to the successes of the firm is its investment in Tesla, Inc. Generation Investment Management identified Tesla as a company that held enormous potential to help catalyze a shift in the global mix from fossil fuels to renewable energy sources through electric vehicles. This particularly illustrates the critical importance of long-term investment strategies that take into account forward-looking sustainability attributes (Generation Investment Management 2020).

2. One of Triodos Bank's success stories is financing towards the wind energy sector. This bank has provided funding to a good number of projects on wind energy spread across Europe, thus growing renewable capacity. By putting emphasis on sustainability in its lending practices, it is proven by Triodos Bank that banks are very important players for promoting sustainable development with high solid financial performance. The bank's experience highlights the potential for financial institutions to drive positive change through targeted lending and investment strategies (Triodos Bank 2020).
3. AXA Investment Managers: AXA Investment Managers is one more example of successful adoption of the principles of sustainable finance. Early in the 2010s, AXA IM started its process of integration of ESG factors into the investment processes within its assets. The company built a corporate scoring system that allowed it to take greater account of companies' ESG performance. This belief in sustainable investment was evident in 2016 when AXA IM decided not to invest in tobacco. From the knowledge of health and economic long-term risks attributed to tobacco products, the investment manager took this step. AXA IM divesting from tobacco moved the group into line with its commitment to sustainability and served to give other investors a case study to consider wider societal impacts from their investment decisions. The firm's experience suggests that aligning investment practices with sustainability principles accomplishes both ethical and financial objectives (AXA IM 2016).

1.4 Challenges and Criticisms

1.4.1 Common Obstacles Faced by the Industry

1. Non-Standardization: One of the prime challenges to sustainable financing is the non-standardization of ESG criteria and reporting. Different organizations and rating agencies, according to them, have used different methods to project the performance of ESG. There are inconsistencies, and it is difficult to compare the efforts made by companies towards sustainability, as Eccles et al. (2012) have stated. The lack of uniformity can confuse investors and delegitimize the credibility of ESG ratings.
2. Data Quality and Availability: The other key challenge is data quality and availability. Most companies either do not divulge comprehensive or reliable data on their sustainability practices or are not really keen to do so—a challenge in making informed decisions on the part of investors. Lack of reporting frameworks further aggravates this condition since companies will tend to disclose favorable information and not anything that is adverse (Chatterji et al. 2016).
3. Short-Termism: Traditional focus of the financial sector on short-term financial performance is likely to be at cross-purposes with long-run sustainability objectives. The need to optimize immediate returns might reduce priorities toward long-term sustainability by investors and firms alike, hereby impeding diffusion of practices in sustainable finance. This can be driven by market pressures, performance benchmarks, and compensation structures (Barton 2011).

1.4.2 Criticisms and Controversies

There is an increased likelihood of greenwashing, which involves exaggeration or lying about environmental and social responsibility practices by firms, being linked to sustainable finance. Such an issue is a common criticism. The main problem with greenwashing is that it misleads the investors and consumers and wears down trust in sustainability claims. High profile cases of greenwashing have propelled doubts relating to the sincerity of corporate sustainability efforts (Delmas and Burbano 2011).

1. Performance Trade-Offs: Critics have argued that sustainable investment could, in fact, lead to the underperformance of traditional investments. There may be some trade-off between financial returns and sustainability goals. Although some literature has claimed that ESG investments can deliver competitive returns, others indicate that results may defer in accordance with market conditions, industrial sectors or, more in general, the criteria used to select ESG firms (Revelli and Viviani 2015).
2. Challenges in ESG Integration: Integration of the ESG factors in investment decisions comes with serious practical challenges. One might face issues related to

ESG considerations in integration with traditional financial metrics. The second issue is concerns about the lack of guidance for ESG integration, sometimes making it bring inconsistent application; furthermore, there has been a debate about the materiality of some ESG factors, which has made the consensus on the impact of ESG factors on financial performance challenging. (Schoenmaker and Schramade 2019).

1.4.3 Responses and Adaptations

1. Setting Standards and Framework: With the weakness of the prevailing reporting standards, various organizations issued frameworks and standards that enhance ESG reporting and assessment. GRI and SASB provide guidance on standardized reporting of such sustainability information. The European Union's SFDR aims at achieving comparability and transparency in ESG information (GRI 2013).
2. Enhanced ESG Data and Analytics: Businesses are tapping into technological and analytical advances to strengthen the gathering and analysis of ESG data. Companies are increasingly leveraging big data, artificial intelligence, and blockchain technology to enable accuracy and reliability in ESG information. Most the third-party providers offering ESG data have expanded their coverage and continue improving their methodologies to drive better and more comprehensive ESG information (Henisz et al. 2019).
3. Long-Term Investment Strategies: The message to be spread is regarding long-term investment strategies, in order to fight the challenge of short-termism. In fact, more and more recent approaches by asset managers and institutional investors have been aligning the investment time frames with sustainability goals. For example, the Principles for Responsible Investment encourage signatories to focus on long-term value creation through embedding ESG factors into processes (PRI 2020).
4. Regulatory and Policy Interventions: Regulatory and policy interventions have played a key role in addressing criticisms and controversies related to sustainable finance. Governments and regulatory bodies are factoring in measures against greenwashing by making provisions for stricter disclosure requirements and punishment against false sustainability claims. Besides, there are policy-related initiatives aimed at promoting sustainable finance, including incentives in terms of tax breaks for green investments and support for ESG integration (TCFD 2017). Challenges and entire criticisms of the sustainable finance industry have a number of responses and adaptations to address those issues. The institutionalization of development standards, along with better ESG data and analytics, the promotion of long-term investment strategies, and regulatory interventions, continues to strengthen credibility and drive the advancement of sustainable finance.

1.5 Future Directions

1.5.1 Emerging Trends and Innovations

1. Technological Integration: Advanced technologies are bound to reshape sustainable finance radically. For instance, blockchain technology avails improved transparency and traceability in ESG reporting and green finance. Blockchain can help in tracing the deployment of sustainable investments and ensure that funds are applied for the purposes for which they were intended. Moreover, AI and big data analytics are being used to externalize vast reams of ESG data that give more insight into companies' sustainability practices and provide more informed investment decisions (Giudici et al. 2020).
2. Climate Risk Assessment: In view of the fact that climate change still poses a virile risk to world GDP, more emphasis is being laid on integrating the climate risk assessment into mainstream financial decision-making; though at an early stage of implementation, TCFD is rapidly becoming a preeminent framework to empower companies to disclose information about climate-related risks and opportunities. This trend is likely to spread, with more and more investors and regulators being involved in a thorough review of the climate risk for companies to understand its financial implications better (TCFD 2017).
3. Green Bonds and Sustainable Finance Instruments: The outlook for exponential growth in the market for green bonds and other instruments of sustainable finance remains strong. So far, green bonds, raising funds specifically for financing projects that are environmentally friendly, have seen heavy issuance in recent years and could turn out to be the major means of financing sustainable development. Examples include blue bonds for the preservation of the seas and transition bonds to help companies change over to green practices, expanding the available instruments in sustainable finance (Climate Bonds Initiative 2020).

1.5.2 Potential Future Developments in Sustainable Finance

1. Enhanced ESG Reporting Standards: The next frontier in sustainable finance lies in more robust and standardized ESG reporting frameworks. Enhanced standards will improve the consistency and comparability of ESG data, which, in turn, allows investors to make more informed decisions (IFRS Foundation 2020).
2. Integration of Biodiversity and Natural Capital: As more people know about the loss of biodiversity and ecosystem degradation, we should expect more interest in incorporating biodiversity and natural capital considerations into financial decision-making. There are massive material risks associated with its loss that financial institutions are just beginning to reckon with by developing frameworks for assessing and mitigating them. The development of financial products and

services that help realize biodiversity conservation and the sustainable use of natural resources is expected to make headway (Deutz et al. 2020).

3. Rise of Social Bonds and Human Capital Investments: The COVID-19 pandemic has emphasized the role of social factors in investment decisions. This is the era where we are likely to see a rise in the issuance of social bonds, which fund projects or with positive social outcomes such as affordable housing, health, and education. Relatedly, human capital investments in areas of workforce development, health, and education are also likely to increase as investors seek to support sustainable and inclusive growth (OECD 2020).
4. Regulatory Developments and Policy Support: The driving force for the growth of sustainable finance is most likely to come from policy support by governments and regulators. The introduction of the European Green Deal and the climate agenda of the Biden Administration in the USA are examples of more policy attention being paid to these issues. Future regulatory developments will include compulsory ESG disclosure, incentives for green investments, along with much more stringent penalties for greenwashing. Such measures will stimulate the implementation of sustainable finance practices across the financial sector (European Commission 2019).
5. Greater Focus on Impact Measurement and Reporting: With the growth in impact investing, there is going to be a much-needed orientation toward measuring. Investors are yearning to see more transparency and accountability about the outcomes their investments achieve. Standardized impact measurement frameworks and tools would enable more effective measurement of whether investments are achieving intended social and environmental outcomes. Such standardization has been prescribed by GIIN only in 2020. From there, it is highly expected that the future of sustainable finance is going to be driven by technological innovation, the parallel improvement of ESG reporting standards, biodiversity, the issuance of social bonds, regulatory support, and improved impact measurement. Growth and maturity in sustainable finance will be driven by trends and developments at large, supporting a transition into a more sustainable and resilient global economy.

1.6 Conclusion

An examination of sustainable finance suggests its deep history and significant impact on contemporary financial practices. From the early investments based on ethics undertaken by religious organizations unto the formation of the Interfaith Center on Corporate Responsibility (ICCR) on to the development of the ESG frameworks, sustainable finance has seen significant development. The development of green bonds and ESG funds, among other sustainable financial instruments, has evolved to create new opportunities for investors trying to combine financial goals with ethical and environmental considerations. Some basic crucial concepts that have contributed to developing this area of sustainable finance include ESG criteria, corporate social responsibility, impact investing, and the triple bottom line: People, planet, profit. Such frameworks and principles have allowed non-financial factors to enter into investors' decision-making processes in order to create more long-term value, with the parallel of ethical accountability.

Therefore, sustainable finance has become a wave of change in the financial sector. It depicts ethically sound and sustainability-compatible financing, one that brings down long-term general risks linked to global challenges in climate mitigation, social inequality, and resource scarcity. Moving money into projects and companies to be able to take the lead in the low-carbon transition for social development and corporate accountability is said. Besides, it underlies that issues of financial performance are interrelated with issues of sustainability. Thus, it challenges the traditional view that considerations of ethics and financial returns exclude each other and flies in the window with potential competitive financial performance of the sustainable investment in generating positive social and environmental impact. It goes without doubt, therefore, that with the growing progress of a sustainable finance approach in this area, it will equally be outstandingly visionary in bringing in the future of finance worldwide. Screen investors, companies, and policy actors then are tasked with the responsibility to pool their efforts jointly towards such a common goal of taking up the gauntlet in innovating transparently and practicing sustainable finance. This success, in the end, according to arguments from sustainable finance, will be a sure contribution that the global goals in sustainability must register, and such will guarantee an equitable, prosperous future for all.

Data Viability Statement NA.

Conflict of Interest There is no conflict of interest.

Funding Statement There is no funding for this research.

References

AXA Investment Managers (2016) AXA Group to divest from tobacco industry assets. Retrieved from https://www.axa-im.com
Barton D (2011) Capitalism for the long term. Harv Bus Rev 89(3):84–91
Bugg-Levine A, Emerson J (2011) Impact investing: transforming how we make money while making a difference. Jossey-Bass
Carroll AB (1991) The pyramid of corporate social responsibility: toward the moral management of organizational stakeholders. Bus Horiz 34(4):39–48
CDP (2020) Carbon disclosure project. Retrieved from https://www.cdp.net
Chatterji AK, Levine DI, Toffel MW (2016) How well do social ratings actually measure corporate social responsibility? J Econ Manage Strategy 18(1):125–169
Climate Bonds Initiative (2020) Green bonds. Retrieved from https://www.climatebonds.net
Cohen R (2014) Impact: reshaping capitalism to drive real change. Ebury Publishing
Delmas MA, Burbano VC (2011) The drivers of greenwashing. Calif Manage Rev 54(1):64–87

Deutz A et al (2020) Financing nature: closing the global biodiversity financing gap. The Paulson Institute, The Nature Conservancy, and the Cornell Atkinson Center for Sustainability. Retrieved from https://www.nature.org

Eccles RG, Ioannou I, Serafeim G (2012) The impact of corporate sustainability on organizational processes and performance. Manage Sci 60(11):2835–2857

Elkington J (1998) Cannibals with forks: the triple bottom line of 21st century business. New Society Publishers

European Commission (2018) Sustainable finance disclosure regulation (SFDR). Retrieved from https://ec.europa.eu/info/business-economy-euro/banking-and-finance/sustainable-finance_en

European Commission (2019) The European green deal. Retrieved from https://ec.europa.eu/info/strategy/priorities-2019-2024/european-green-deal_en

Friede G, Busch T, Bassen A (2015) ESG and financial performance: aggregated evidence from more than 2000 empirical studies. J Sustain Fin Investment 5(4):210–233

Generation Investment Management (2020) Annual sustainability report. Retrieved from https://www.generationim.com

Giudici G, Milne A, Vinogradov D (2020) Cryptocurrencies: market analysis and perspectives. J Industr Bus Econ 47(1):1–18

Global Impact Investing Network (GIIN) (2020) Annual impact investor survey. Retrieved from https://thegiin.org

Global Reporting Initiative (GRI) (2013) Sustainability reporting guidelines. Retrieved from https://www.globalreporting.org

Gond J-P, Kang N, Moon J (2012) The government of self-regulation: on the comparative dynamics of corporate social responsibility. Econ Soc 40(4):640–671

Henisz W, Koller T, Nuttall R (2019) Five ways that ESG creates value. McKinsey Quarterly. Retrieved from https://www.mckinsey.com/business-functions/strategy-and-corporate-finance/our-insights/five-ways-that-esg-creates-value

ICMA (2017) Social bond principles. Retrieved from https://www.icmagroup.org

IFRS Foundation (2020) Consultation paper on sustainability reporting. Retrieved from https://www.ifrs.org

Kurtz L (2008) Socially responsible investment and shareholder activism. In: Crane A, Matten D, McWilliams A, Moon J, Siegel D (eds) The Oxford handbook of corporate social responsibility. Oxford University Press, pp 249–279

Loan Market Association (2019) Sustainability linked loan principles. Retrieved from https://www.lma.eu.com

Morningstar (2021) Global sustainable fund flows. Retrieved from https://www.morningstar.com

Norman W, MacDonald C (2004) Getting to the bottom of "triple bottom line." Bus Ethics Q 14(2):243–262

O'Donohoe N, Leijonhufvud C, Saltuk Y, Bugg-Levine A, Brandenburg M (2010) Impact investments: an emerging asset class. J.P. Morgan

OECD (2020) Social impact investment 2019: the impact imperative for sustainable development. OECD Publishing. Retrieved from https://www.oecd.org

Pax World (2020) Pax World funds. Retrieved from https://www.paxworld.com

Porter ME, Kramer MR (2006) Strategy and society: the link between competitive advantage and corporate social responsibility. Harv Bus Rev 84(12):78–92

Principles for Responsible Investment (PRI) (2020) Annual report 2020. Retrieved from https://www.unpri.org

Renneboog L, Ter Horst J, Zhang C (2008) Socially responsible investments: institutional aspects, performance, and investor behavior. J Bank Finance 32(9):1723–1742

Revelli C, Viviani JL (2015) Financial performance of socially responsible investing (SRI): what have we learned? A meta-analysis. Bus Ethics: Eur Rev 24(2):158–185

Schoenmaker D, Schramade W (2019) Principles of sustainable finance. Oxford University Press

SEC (2021) SEC announces enforcement task force focused on climate and ESG issues. Retrieved from https://www.sec.gov/news/press-release/2021-42

Sievänen R, Rita H, Scholtens B (2013) The drivers of responsible investment: the case of European pension funds. J Bus Ethics 117(1):137–151
Slaper TF, Hall TJ (2011) The triple bottom line: what is it and how does it work? Indiana Bus Rev 86(1):4–8
Sparkes R (2002) Socially responsible investment: a global revolution. John Wiley and Sons
Storebrand (2021) Storebrand annual report 2020. Retrieved from https://www.storebrand.com
Task Force on Climate-related Financial Disclosures (TCFD) (2017) Final report: recommendations of the task force on climate-related financial disclosures. Retrieved from https://www.fsb-tcfd.org
Triodos Bank (2020) Annual report 2019. Retrieved from https://www.triodos.com

Chapter 2
Sustainable Investment for Renewable Energy in Indian Context

Sunita Jatav and Neeraj Gupta

Abstract Renewable energy is abundance and meticulous use in India. Renewable resources are readily available and offer clean sustainable energy. Fossil fuels reserves are deteriorating the demand and energy is increasing drastically. In response to limited fossil fuel, the global community is focusing on renewable energy resources. Renewable energy sources generate 24% of global electricity, but their capacity to meet future demand is restricted. Since India is a fast-growing economy, it needs a larger energy supply than it currently consumes, for which renewable energy is a feasible option to meet the demand. India intends to produce 200 GW of renewable energy by 2030. The expansion of renewable energy has constraints. This book chapter discusses renewable development difficulties and offers solutions. The study measures the successes and challenges of India's renewable energy development and suggests an investment strategy for continued expansion.

Keywords Barriers · Investment · Renewable energy · Strategies · Sustainable

2.1 Introduction

Renewable energy sources and technologies offer viable solutions to the energy challenges in developing nations. Wind, solar, geothermal, ocean, biomass, and fuel cell technology can solve India's energy problems. As a rapidly developing economy, India needs a significantly larger energy supply than it currently consumes, and renewable energy is a feasible option to meet this demand. About 33% of India's primary energy comes from renewable sources. India is methodically implementing renewable energy projects to reduce carbon emissions and improve air quality (Kumar et al. 2010). India ranks fourth globally in total CO_2 emissions, which have

S. Jatav (✉) · N. Gupta
N. L. Dalmia Institute of Management Studies and Research, Mumbai, India
e-mail: sunita.jatav@nldalmia.edu.in

N. Gupta
e-mail: neeraj.gupta@nldalmia.edu.in

R. Kumar (ed.), *Sustainable Finance*, Sustainability Solutions,
https://doi.org/10.1007/978-3-031-97962-0_2

increased significantly since the 1990s (World Bank). India must urgently advance renewable energy initiatives and diminish fossil fuel usage. In the 1990s, several small hydropower facilities commenced electricity generation in India. Nevertheless, considering the technological advancements of that era, the output of the stations was significantly limited (Thomas 2015). Since that period, hydropower has developed gradually yet consistently. India began renewable energy research in the 1960s. To harness wind energy, the National Aeronautical Laboratory built a typical multi-vane windmill in the mid-1960s (Zeng et al. 2017). In the subsequent decade, various types of windmills were constructed. The initial advancements in wind energy were aimed at irrigation applications (Zeng et al. 2017). In the 1970s and 1980s, there were national solar PV and thermal programs (Bhattacharya and Jana 2009). Building upon these experiences, India subsequently integrated solar energy into everyday life. By 1995, 700 photovoltaic pumps, with 26,000 residential light units, and more solar energy systems (Zeng et al. 2017) were installed. Recently the established wind farms have commenced power generation in the Indian market. Globally renewable energy is considered as effective way to reduce CO_2 emissions (Mitigation 2011). India is one of the famous nations with over 1.4 billion residents and significantly contributes to global emissions and climate change due to its carbon intensity (Elavarasan et al. 2020). In 2021, India released around 3.9 billion tons of CO_2-equivalent greenhouse gases, positioning it as the third-largest emitter, following China and the USA. Nonetheless, its per capita greenhouse gas emissions were 2.8 CO_2-corresponding tons, in contrast to the global standard of 6.91 and 17.51 in the USA (Hussain and Dill 2023). The World Energy Council forecasts that the world's requirement for energy will arise in the year 2030. India, a major coal consumer, imports expensive fossil fuel, with approximately 74% of its energy necessity met by natural resources (World Energy 2013). The CMIE says that India traded 171 million tons of coal in 2013–14, 215 million in 2014–15, 207 million in 2015–16, 195 million in 2016–17, and 213 million in 2017–18 (Blondeel and Van de Graaf 2018). Online education reduces carbon footprint (Mohanty et al. 2024) suggests that online learning can reduce greenhouse gas (GHG) emissions and environmental impact. The research examines the achievements and barriers in India's renewable resources development and outlines an effective investment strategy for further advancement.

2.2 Literature Review

2.2.1 India's Demand for Primary Energy Consumption

Population increase significantly impacts energy demand in the fastest-growing nation. It makes up 17.74% of the global population and grows 1.18% annually. Population predictions for India include 1.383 billion in 2020, 1.512 billion in 2030, 1.605 billion in 2040, and 1.658 billion in 2050 (Kumar and Majid, 2020). India's population growth is substantial relative to the world average, with many states possessing populations as per whole nations (Avesh et al. 2024). India's anticipated energy spending growth is substantial by the year 2024. This increase will be primarily driven by coal, and thereafter, it will be fueled by renewable resources. India needs 256 MT of renewable energy by 2040, up 12% annually (World Meters 2019; Kumar et al. 2020) (Tables 2.1 and 2.2).

India's economy is expanding at a rapid growth rate, with scarcity of energy. In current scenario, India has robust economic progress, for that country's need significant and additional energy sources to satisfy this demand (Kumar et al. 2024b). During this time, the country is struggling with the challenge of achieving sustainable development because of the cumulative residents and the decline of environment quality (Arto 2016). It is anticipated that the gap between the demand and supply of power would widen in coming years. According to International Renewable Energy Agency (IREA), India possesses significant potential to fulfill its complete demand. It is possible that the country will increase the total production of renewable energy by the 2030. Table 2.3 delineates the energy source position of the nation from 2009–2010 to 2023–2024 (up to October 2024). In 2024, energy demand reached 266,951 GWh, while availability was 266,360 GWh, resulting in a deficit of − 0.2% (Ministry of Power 2023).

2.2.2 *Types of Energy in India*

Solar Energy

MNRE's National Solar Mission increased grid-connected solar power facility targets from 20 to 100 GW by 2021–2022. It was 6 MW tell by 2008–2009. The "Made in India" campaign increased local manufacturing and solar installation capability. India stands fifth globally in solar installed capacity. By December 31, 2018, solar energy reached 25,213 MW, with an additional 22.8 GW under tender or implementation (Kumar and Majid 2020). The MNRE will auction the remaining capacity each year from 2018 to 2020 to reach 100 GW by March 2020, giving two years for project completion. Competitive reverse e-auctions will set tariffs, leading in significant cost reductions, with the lowest cost in 2018 being INR 2.45 per kWh, down from INR 18 in 2010. More than 100,000 acres have been entitled for solar parks, with 75,000 acres secured (Kumar and Majid 2020).

Wind Energy

India's installed wind power was 35,138.15 megawatt on December 31, 2018, compared to 60 gigawatt by 2022. India ranks fourth in wind power capacity, with 9.4 GW tendered or under construction. For 2018–2019 and 2019–2020, the MNRE will auction 10 GW of wind energy capacity every year, aiming to build 60 GW by March 2020 (Table 2.3). The nation has 302 GW of 100-m-high wind energy capacity. The administration of tariffs has transitioned from a feed-in tariff system to

Table 2.1 World-wise energy consumption between (1990–2040) (BP Energy Outlook 2018; Kumar J and Majid 2020)

Region	Consumption (million tons oil equivalent)											Changes (% per annum)	
	1990	1995	2000	2005	2010	2016	2020	2025	2030	2035	2040	1990–2016	2016–2040
USA	1965	2118	2311	2348	2285	22,723	2333	2345	2342	2324	2298	0.61	0.00
Brazil	125	157	187	212	269	299	331	379	418	452	478	3.41	2.00
EU	1671	1662	1733	1818	1755	1643	1668	1624	1571	1512	1461	– 0.11	– 0.51
Russia	866	661	621	648	674	675	712	721	724	723	717	– 2.00	0.31
Middle East	265	352	424	566	735	896	981	1086	1188	1288	1383	4.80	1.80
Africa	223	245	275	328	388	441	508	602	711	841	1003	2.70	3.50
China	682	888	1009	1801	2492	3054	3388	3754	4018	4208	4318	5.90	1.50
India	196	252	317	395	538	725	881	1119	1366	1625	1922	5.20	4.20

Source http://www.prayaspune.org

Table 2.2 India's primary energy usage (including of renewable energy) from 1990 to 2024 (BP Energy Outlook 2018; Kumar J and Majid 2020)

	Level (Mtoe)												
	1990	1995	2000	2005	2010	2015	2020	2025	2030	2035	2040	1990–2016 (%)	2016–2040 (%)
Total	196	252	317	393	536	723	881	1119	1365	1625	1922	5.21	4.21
Oil (Mb/dI)	59	76	107	123	156	213	252	309	358	418	486	5.11	3.51
Gas (Bcf/dI)	12	18	25	33	55	46	58	73	88	107	129	5.61	4.50
Coal	111	141	165	212	291	413	486	594	711	825	956	5.20	3.60
Nuclear	1.1	2.1	4.1	4.1	5.1	9.1	12	17	28	36	45	7.10	7.00
Hydro	16	18	16	23	26	28	37	44	48	51	53	2.60	2.50
Renewables	0	0	1.1	2.1	7.1	18	42	87	132	192	257	35.11	12.00

Source http://www.prayaspune.org

Table 2.3 Capacity shares for renewable energy facilities (installed)

Sources (GW)	2002	2007	2012	2017	2022	2027	2032
Thermal	74.5	87	131.7	218.2	296.2	389.2	552.3
Renewables	3.41	10.21	24.91	57.13	176	250.41	350.31
Nuclear	2.71	3.91	4.71	6.79	11.83	19.11	23.21
Large hydropower	26.21	34.61	38.91	44.47	60.91	67.11	74.51
Total	106.8	134.8	200.2	326.87	543.85	725.19	1000.5
Percentage renewable energy capacity shares	3.1	8.1	12.1	18.1	32.1	34	35

Source http://www.prayaspune.org

a request approach, with the ministry releasing instructions for tariff-based competitive bidding on December 8, 2017. The transparent bidding process has lowered wind power tariffs to their minimum levels. The wind sector has established a strong ecosystem with operational capabilities and a manufacturing foundation, providing advanced wind turbine technologies (Kumar et al. 2024a). All major global wind power players are present in India, with over 12 companies producing more than 24 wind turbine models. India sells wind turbines and components to the USA, Europe, Australia, Brazil, and other Asian countries, with strong indigenous manufacturers producing 70–80% (Kumar and Majid 2020).

2.3 Research Methodology

It provides a comprehensive framework to understand the broader environment in which the country operates. The study utilizes qualitative research methods to achieve its objectives, focusing on the successes and challenges of India's renewable energy development and suggests an investment strategy for continued expansion. This study will focus on the necessity of implementing sustainable practices while specifically examining the obstacles to commercializing renewable energy in India. The study will also propose financial strategies and initiatives to support the renewable energy sector.

2.4 Interpretation and Result

2.4.1 Need of Sustainability into Practice

The principles of "sustainability" and "sustainable development" are essential in environmental policy and international development, yet their understanding is often impeded by the absence of a clear, universally accepted definition. However, to guide decision-making and policy formulation, these concepts must be applied across various academic fields. This diagram aids in understanding the objectives and interrelationships within these areas. Developing a sustainable framework for energy supply and utilization operationalizes the concept of sustainability. It is necessary to have an awareness of the connections that exist between energy, socioeconomic growth, the environment, and socioeconomic security to develop effective solutions. A sustainable energy system could use more renewable energy, more efficient fossil fuel production and use, and fuel substitution from high to low and no-carbon fuels. The adoption and implementation of sustainable energy is essential for growth. Over the next two decades, a rise in renewable energy usage is anticipated, primarily due to the low cost of fossil fuels hindering renewable technology advancement. Accelerating renewable energy growth and enhancing its contribution to current energy mixes require rigorous application of relevant theories and advance technologies.

2.4.2 Barriers to the Commercialization of Renewable Energy in India

Renewable energy strategy is promptly advancing to reinforce renewable technologies, aiming to utilize previously untapped renewable resources (Kumar and Majid 2020). Expanding renewable capacity poses significant challenges. The MNRE has taken steps to foster a renewable environment and identify barriers, but these measures have not sufficiently aided the advancement of renewable technologies (IDFC 2010). Thus, auditing the current renewable landscape is necessary for ongoing development. Certain barriers may be specific to policy, technology and region. General barriers affect all renewable resources, whereas source-specific barriers pertain to individual sources. Various studies (IDFC 2010; Khare et al. 2013; Nesamalar et al. 2017; Flamos 2010; Uddin et al. 2010; Osorio et al. 2005) address these barriers:

Regularity and policy barriers

- Indian government needs to encourage growth of renewable energy technologies, due to huge requirements.
- India lacks a comprehensive policy declaration on renewable energy technology.
- Current improvement of plan does not equivalent with current policies

- At the state level, the policy framework is inadequate, leading to significant uncertainty in regulations for renewable technology investment.
- The legal framework is insufficient for the renewable purchase obligation (RPO), which applies to a range of different resources of renewable energy generation (Kumar and Pal 2020).

Institutional barriers

- Due to lack of skill among various ministries, institutes, and agencies in implementing renewable energy policies in India, which undermines investor confidence in renewable energy developments.
- Gap in securing approval for projects via undertaking incurs penalties for the maker. Consequently, the authorization system is inadequate in all states except Punjab, which has established a 60-day deadline for approvals.
- Most states lack adequate planning of pre-feasibility reports (PFR) by organizations.
- The institutions accountable for training manpower for new renewable infrastructure are inadequate and too few.
- The malfunctioning of the existing institutions and develop renewable infrastructure.
- Most institutions lack dedicated research centers for renewable infrastructure development.
- Insufficient organizations and laboratories in the country exist to set standards or certifications for renewable technology quality and appropriateness (Kumar and Majid 2020).

Economic barriers

- The resources for renewable developments in India are not released promptly.
- Financing is a significant challenge for renewable projects. Despite low fuel and operational costs, renewables incur high capital expenses, which can hinder funding efforts.
- Subsidies are not provided for renewable energy shortages or cost reductions.
- Renewable projects are smaller and offer lower returns than non-renewable ones, making them less attractive to investors.
- Renewable developers are small and newly commercialized, making lenders reluctant to finance their projects.
- Local and small developers encounter financial challenges due to the high investment required for renewable projects. Small projects carry the same financial and business hazards as larger ones, deterring stakeholders from getting involved.
- Small renewable projects face overly high yielding costs and bureaucracy compared to their finance needs.
- Renewable energy underperformance hinders funding for initiatives.
- Financial organizations lack proficiency in renewable energy, creating a barrier to financing such projects (Kumar and Majid 2020).

Market-related barriers

- The cost-plus calculation for renewable energy in India is inefficient, ensuring charge improvement for developments.
- The high expense of renewable energy is due to overlooking environmental expenses, obscuring clean energy's environmental benefits.
- The current subsidy system favors conventional resources, making renewable energy more expensive due to this inequitable subsidy structure in the power market.
- India exclusively supports government-driven renewables, neglecting loan and finance markets with significant potential for development.
- The primary challenges in renewable energy development are grid integration and infrastructure availability. Additionally, seasonal variability and the typically small scale of renewable projects hinder progress, as extending transmission lines for these projects is often not cost-effective.
- Transmission utilities are responsible for delivering renewable power but lack knowledge of transmission of non-renewable power.

Technical barriers

- Renewable technologies in India are still progressing, resulting in high technological risks. The uncertainties surrounding these technologies mean their expected performance in the Indian context is suboptimal.
- India has limited R&D centers for renewable technologies.
- The nation's production enterprises are constrained to relatively modest operations and mostly concentrate on the replication of existing technology.
- The country relies on global suppliers for the resources and technologies utilized in renewable energy.

Information barriers

- Lack of public information hinders the extensive acceptance of renewable technologies in India.
- There is a lack of consistent programs aimed at promoting awareness of technologies.
- The display, training, operation, and maintenance of renewable technology are all areas in which there is a dearth of trained workers.
- A deficiency exists in the comprehension of practical issues and the general familiarity of emerging knowledges.

Social barriers

- Renewable energy is not widely adopted due to the easy availability and lower cost of conventional energy sources.
- India's population predominantly resides in villages, with about 70% living in rural areas. Villagers are generally unfamiliar with renewable energy knowledges. The national literacy rate is around 72%, varying by state, which suggests limited comprehension of new technologies.

- India, with a population density of 382 people per square kilometer, faces challenges in meeting the high land demand for renewable energy. Since 60–70% of the population relies on agriculture, most land is dedicated to this sector, making land availability scarce.
- The intermittent nature of renewable energy necessitates storage systems, which raise the overall cost and deter people from adopting this technology.
- The reliance of renewable technology on climate conditions diminishes its appeal within society. The distance of Indian villages from urban centers results in inadequate customer support for renewable technology.
- There exists a dearth of information in India concerning the environmental advantages of renewable technology, and unfavorable attitudes around these technologies operate as a hurdle
- Transportation infrastructure in rural India is inadequate, presenting a considerable obstacle to the advancement of renewable technology.

Environmental barriers

- Deforestation negatively impacts the environment by reducing the absorption of harmful gases by trees, contributing to global warming. Additionally, renewable energy development can harm wildlife.
- The renewable system necessitates extensive land, leading to deforestation for construction, thus contributing to forest loss.
- Deforestation exerts a detrimental impact on climatic conditions. For instance, the hydrological cycle is disrupted as the processes involved in the water cycle are impeded due to deforestation.
- Deforestation alters heavy rainfall, adversely impacting agriculture. Renewable systems require extensive land, significantly affecting soil erosion and drainage channels.

Energy-related barriers (solar, wind, and small hydropower)

- Solar energy is unavailable during the night or on cloudy days, necessitating a backup for continuous electricity supply, which adds complexity to solar technology.
- There is a need for technological advancement in the energy storage system as well as the large thermal losses that occur in the solar thermal. Solar thermal applications in various regions, like Rajasthan, here solar energy has tremendous potential, are primarily concerned with the scarcity of water. This is the fundamental worry.
- Solar energy technology is less efficient than regular energy, making economic viability challenging.
- The overall efficiency of photovoltaic systems in residential applications is diminished due to the lack of solar tracking and the natural degradation of photovoltaic cells, which includes issues such as cracking, discoloration, snow cover, hotspots, and dust accumulation.

- Over time, solar panels accumulate dust, necessitating regular cleaning to extend their lifespan and maintain efficiency; however, no cleaning system is currently available in the country.
- The short lifespan of the storage device, such as the battery, necessitates frequent replacements, raising overall costs. While solar PV product prices are falling in the international market, they remain high in the Indian market.
- Installing a utility-scale solar plant necessitates a large land area, potentially disrupting existing land uses. This extensive land use leads to grading and clearing, which can cause soil erosion and compaction.
- Wind development policies and subsidies in many Indian states are short-term. For instance, in Madhya Pradesh, if wind energy generation meets or exceeds 70% of the declared amount, the policy exempts electricity payments for only five years.
- The Indian government's unequal support for wind power growth is evident. It offers 80% tax depreciation for companies investing in wind development. Consequently, foreign investors show little interest as they have no income tax to save.
- The country's equipment suppliers, who also develop wind projects, sell these ready-made projects at a premium, making them more expensive for buyers.
- The geographical setting influences how people perceive the landscape and assess visual impact, with natural features serving as a significant obstacle to wind farm development.
- State government utilities are authorized to prepare PFR for small hydropower project, but it is not a high priority area for the government, causing delays in project execution and allocation.
- The primary factor influencing SHP development is water availability. Given that India's economy relies heavily on agriculture, water consumption is prioritized for this sector, making it a significant obstacle to advancing SHP technology.
- The country's hydrological and geological conditions cause significant delays in SHP project completion, often taking 8–12 years, which deters investor interest.
- The lack of practical and expert civil contractors for SHP developments is a significant barrier. Additionally, strong labor unions and local law and order issues in many states can further impede development.

2.4.3 Sustainable Financial Strategy for Renewable Energy

Financing for renewable energy, especially wind and solar, has entered mainstream investment with substantial private financial flows. Assets are often managed by private equity, pension, sovereign wealth funds etc. The establishment of Solar Energy Corporation of India (SECI) has notably improved sector credit quality. However, tripling or quadrupling the current RE deployment rate will require expanding financing options and frameworks. Innovative instruments like Infrastructure Investment Trusts (InVITs), new private equity funds, green bonds, and blended financial instruments will be essential. These innovations are also crucial for newer technologies like hydrogen, floating solar, and seaward wind, which may not yet be fully bankable due to factors like unproven technology, higher risks, and initial costs.

To ensure sustained progress, the government has set a renewable purchase obligation (RPO) target of 43.33% by 2030, with a specific RPO for distributed renewable energy. This commitment outlines a clear path for meeting renewable energy targets and advancing cleaner energy sources. New initiatives include the Development of Solar Parks, Pradhan Mantri Kisan Urja Suraksha evam Utthaan Mahabhiyan Yojana (PM KUSUM), the Grid Connected Solar Rooftop Programme, PM-Surya Ghar: Muft Bijli Yojana, the National Framework for promoting energy storage systems, and the Viability Gap Funding (VGF) scheme for 4GWh of energy storage systems. Since energy storage devices generate no direct income, VGF helps bridge the gap between their deployment cost and economic viability. VGF can cover up to 40 percent of the capital cost of Battery Energy Storage Systems (BESS), making such projects more attractive for businesses by offsetting initial expenses (mnre.gov.in; powermin.gov.in).

The MNRE requires Renewable Energy Implementation Agencies (REIAs) to bid annually to increase India's renewable energy capacity. This plan solicits proposals for 50 gigawatt of renewable energy annually from 2023–24 to 2027–28, with a minimum of 10 GW for wind power (IREDA 2023). The GOI has taken steps to boost solar, wind, and Green Hydrogen production. As part of a significant initiative, the Inter-State Transmission System solar and wind power projects slated for commission before June 30, 2025, will be waived, and subsequent graded rates will be implemented to decrease financial barriers and encourage the expansion of the industry (Press Information Bureau 2024).

The Ministry of Power allotted Rs. 20,502.1 Cr in 2024–25, up 16% from the revised 2023–24 projection (Table 2.4). Five percent goes to capital expenditure. The Revamped Distribution Sector Scheme (RDSS), introduced in 2021 in the direction of facilitate distribution companies improve financial and operational performance, receives 61% of total spending. Prepaid smart meter installation help is a major RDSS component. Other important allocations include aiding central public sector firms with power projects (which accounts for fifteen percent of the allocation) and enhancing power infrastructures (twelve percent), primarily transmission networks. The MNRE received Rs. 19,100 crore in 2024–25, up 143% over the revised forecast for 2023–24. In February 2024, the PM-Surya Ghar Muft Bijli Yojana gave households financial aid for rooftop solar installations, driving this growth.

In the last ten years, MNRE's actual spending has generally fallen short of budget projections, with the exceptions of the years 2021–22 and 2022–23. For the fiscal year 2023–24, updated projections suggest that the total expenditure of MNRE will fall short of the budget by 23%. The PM KUSUM scheme, aimed at solarizing agricultural pumps, shows a 45% underutilization rate in 2023–24. Furthermore, expenditures on grid-connected wind power initiatives for the 2023–24 period are anticipated to fall 25% short of the allocated budget (see Fig. 2.1).

Table 2.4 Power and renewable energy allocation (Rs. Cr)

Head	2022–23	2023–24	2023–24	% Change 2023–24 from 23–24 BE to RE	2024–25	% Change from 23–24 RE to 24–25 BE	Share in ministry budget in 24–25
	Actuals	BE	RE		BE		
Ministry of new and renewable energy of which	7563	10,222	7848	− 23	19,100	143	
Solar power grid	4280	4970	4757	− 4	8500	79	45
Pradhan Mantri Surya Ghar Muft Bijili Yojna	0	0	0	–	6250	–	33
PM KUSUM	1325	1996	1100	− 45	1496	36	8
Wind power grid	1413	1214	916	− 25	800	− 13	4
Green energy corridor	250	500	434	− 13	600	38	3
Indian green hydrogen mission	0	297	100	− 66	600	500	3

Budget Estimates (BE); Revised Estimates (RE)
Source Union Budget 2024–25

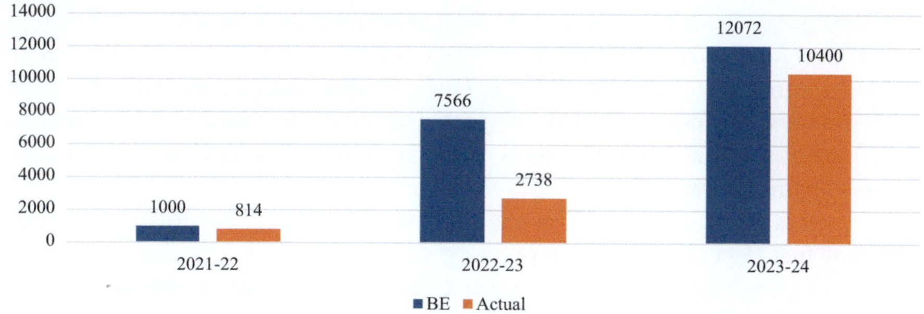

Fig. 2.1 Revamped distribution sector scheme utilised fund (in Rs. crore) (Powermin.gov.in; DFG Analysis 2024–25)

Discoms must get a certain percentage of their electricity from renewable sources. The renewable purchase obligation (RPO) promotes renewable energy demand. 15 states were successful in meeting their RPO in 2022–23. Distribution businesses consider renewable energy expensive, with integration costs, according to the Union Ministry of Power (2022). According to a notification issued in October 2023, the RPO will be 29.9% for the 2024–25 fiscal year, and it will increase to 43.3% by the 2029–30 fiscal year.

2.5 Upcoming Initiatives for Development Strategy for Renewable Energy

To increase renewable energy demand, the renewable purchase obligation (RPO) mandates discoms to buy a minimum percentage of their electricity from renewable sources. Table 2.5 of the annexure shows 15 states met their RPO in 2022–23. Distribution businesses view renewable energy as expensive and require integration costs, according to the Union Ministry of Power (2022) (DFG Analysis 2024–25: Power and New and Renewable Energy). An October 2023 notification sets the RPO at 29.9% for 2024–25, increasing to 43.3% by 2029–30:

1. Madhya Pradesh envisions becoming a renewable energy hub, targeting 50% of its power mix from renewables by 2030, with milestones of 20% by FY 2024 and 30% by FY 2027. The state plans ₹15,000 crore investment in generation by 2024, rising to ₹50,000 crore by 2027, and ₹10,000 crore in equipment manufacturing. Capacity goals include 36,000 MW by 2035, with storage targets of 12,000 MW. Incentives feature tax exemptions, land rebates, and benefits for storage, green hydrogen, and biofuels. Additional measures include Green Zones, skill development, and R&D promotion in technical institutes.
2. Uttar Pradesh accelerating solar power deployment to achieve India's ambitious solar energy capacity goals: 22,000 MW of solar projects by 2026–27. By 2030, UP aims for over 20% RE in its energy mix. Incentives include stamp duty exemption, rooftop solar subsidies, Priority Open Access, and support mechanisms for consumers, businesses, and developers. Initiatives also include plug-and-play solar parks, solarizing agricultural electrical feeders, and special incentives for small hydro projects.
3. Gujarat is the state with RE potential, enhancing energy security by reducing fossil fuel dependency, and promoting decentralized RE generation. Achieve 50% of growing electric power connected power beginning non-fossil fossil fuel by 2030. Plan for an additional 34.5 GW of RE generation capacity. Implement solar park policies, capital subsidies for rooftop solar, Priority Open Access for RE developers, and encourage investment, employment, and skill development in the RE sector. Increase energy awareness.

Table 2.5 Compliance With Renewable Purchase Obligation in 2022–23

State	Target				Achievement			
	Wind	Hydro	Other	Total	Wind	Hydro	Other	Total
Andhra Pradesh	0.81%	0.35%	23.44%	24.61%	3.30%	0.00%	25.10%	28.40%
Arunachal Pradesh	0.81%	0.35%	23.44%	24.61%	0.00%	3.50%	15.30	18.80%
Assam	0.81%	0.35%	23.44%	24.61%	0.00%	2.70%	20.40%	18.80%
Bihar	0.81%	0.35%	23.44%	24.61%	0.10%	0.10%	15.90%	16.10%
Chhattisgarh	0.81%	0.35%	23.44%	24.61%	0.00%	1.20%	11.90%	13.20%
Delhi	0.81%	0.35%	23.44%	24.61%	0.80%	0.60%	23.80%	24.40%
Goa	0.81%	0.35%	23.44%	24.61%	3.50%	0.40%	16.10%	17.30%
Gujrat	0.81%	0.35%	23.44%	24.61%	0.00%	0.20%	16.50%	20.20%
Haryana	0.81%	0.35%	23.44%	24.61%	0.00%	0.90%	20.80%	21.70%
Himachal Pradesh	0.81%	0.35%	23.44%	24.61%	0.00%	2.70%	76.00%	78.70%
Jammu & Kashmir & UT	0.81%	0.35%	23.44%	24.61%	0.00%	0.00%	56.90%	56.90%
Jharkhand	0.81%	0.35%	23.44%	24.61%	6.40%	0.00%	30.00%	30.00%
Karnataka	0.81%	0.35%	23.44%	24.61%	0.00%	0.00%	40.30%	46.70%
Kerala	0.81%	0.35%	23.44%	24.61%	0.10%	0.00%	36.30%	36.30%
Madhya Pradesh	0.81%	0.35%	23.44%	24.61%	0.00%	0.00%	22.30%	22.40%
Maharashtra	0.81%	0.35%	23.44%	24.61%	0.00%	0.00%	17.10%	17.10%
Manipur	0.81%	0.35%	23.44%	24.61%	0.00%	0.20%	33.80%	34.00%
Meghalaya	0.81%	0.35%	23.44%	24.61%	0.30%	3.20%	56.70%	56.90%
Mizoram	0.81%	0.35%	23.44%	24.61%	0.00%	0.40%	42.10%	42.80%
Nagaland	0.81%	0.35%	23.44%	24.61%	0.10%	5.20%	31.30%	36.50%
Odisha	0.81%	0.35%	23.44%	24.61%	0.00%	0.10%	25.30%	25.50%
Puducherry	0.81%	0.35%lePara>	23.44%	24.61%	0.20%	0.00%	6.60%	6.60%
Punjab	0.81%	0.35%	23.44%	24.61%	0.00%	0.00%	27.30%	27.50%
Rajasthan	0.81%	0.35%	23.44%	24.61%	0.00%	0.00%	18.30%	18.30%
Sikkim	0.81%	0.35%	23.44%	24.61%	5.80%	3.70%	84.70%	88.40%
Tamil Nadu	0.81%	0.35%	23.44%	24.61%	0.00%	0.00%	19.70%	25.50%
Telangana	0.81%	0.35%	23.44%	24.61%	0.00%	0.00%	20.20%	20.20%
Tripura	0.81%	0.35%	23.44%	24.61%	0.00%	0.10%	13.50%	13.60%
Uttar Pradesh	0.81%	0.35%	23.44%	24.61%	0.00%	0.40%	14.80%	15.20%

(continued)

Table 2.5 (continued)

State	Target				Achievement			
	Wind	Hydro	Other	Total	Wind	Hydro	Other	Total
Uttarakhand	0.81%	0.35%	23.44%	24.61%	0.00%	3.00%	57.40%	60.40%
West Bengal	0.81%	0.35%	23.44%	24.61%	0.10%	0.00%	15.90%	16.00%

Source Starred Question No. 122, Rajya Sabha, Ministry of Power, August 1, 2023; PRS

1 Entry No. 38, List III- Concurrent List, Seventh Schedule, Constitution of India, https://cdnbbsr.s3waas.gov.in/s380537a945c7aaa788ccfcdf1b99b5d8f/uploads/2024/07/20240716890312078.pdf

2 "About Ministry", Website of Ministry of Power as accessed on July 29, 2024, https://powermin.gov.in/en/content/about-ministry

3 Annual Report 2022-23, Ministry of New and Renewable Energy, https://cdnbbsr.s3waas.gov.in/s3716e1b8c6cd17b771da77391355749f3/uploads/2023/08/2023080211.pdf

4 Electricity Act, 2003, https://cercind.gov.in/Act-with-amendment.pdf

5 Installed Capacity Report - June 2024, Central Electricity Authority, https://cea.nic.in/wp-content/uploads/installed/2024/06/IC_June_2024_allocation_wise.pdf

6 Report on Performance of Power Utilities 2022-23, Power Finance Corporation, https://pfcindia.com/ensite/DocumentRepository/ckfinder/files/Operations/Performance_Reports_of_State_Power_Utilities/Report%20Database%202022-23%20-%20updated%20up%20to%20April%2024EntityApr.pdf

7 Demand No. 71, Expenditure Budget, Ministry of New and Renewable Energy, Union Budget 2024–25, https://www.indiabudget.gov.in/doc/eb/sbe71.pdf

8 Demand No. 79, Expenditure Budget, Ministry of Power, Union Budget 2024-25, https://www.indiabudget.gov.in/doc/eb/sbe79.pdf

9 Budget Speech, Union Budget 2024–25, https://www.indiabudget.gov.in/doc/Budget_Speech.pdf

10 "Cabinet approves Revamped Distribution Sector Scheme: A Reforms based and Results linked Scheme", Press Information Bureau, Union Cabinet, June 30, 2021, https://pib.gov.in/PressReleasePage.aspx?PRID=1731473

4. Karnataka is developing the state as a hub of renewable energy generation energy with target 10 GW of additional RE projects by 2022–27.GOI has introduced stamp duty exemptions, RPO enforcement, banking of power along with strategy to attract investments and tap RE potential. The Renewable Purchase Obligation is scheduled to rise significantly. As per an October 2023 announcement, it will be 29.9% for the 2024–25 financial year, and this figure is set to increase to 43.3% by the 2029–30 financial year.
5. Rajasthan has aims to promote renewable energy integration into the state's power grid and develop 90 GW renewable energy capacity by 2030 and state government has target of 90,000 MW renewable power projects by 2029–30. Government support is happening with 50% exemption on transmission charges, capital subsidy for solar power projects, and land lease policy for solar parks.
6. India has enacted multiple decarbonization and diversification initiatives, targeting net zero emissions by 2070. Recently, there has been an expansion in investing in solar and wind energy, coupled with initiatives to bolster domestic renewable energy supply chains. The 2020 Production Linked Incentives system encourages domestic production of solar panels, batteries, and other clean energy apparatus. Furthermore, India upholds a longstanding energy efficiency initiative and has introduced a hydrogen strategy aimed at domestic electrolyzer

manufacturing and low-carbon hydrogen production (World Energy Investment 2024).

7. Recent initiatives have markedly enhanced Indian renewable energy investment, attaining USD 68 billion in 2023, representing an approximate 40% increase from the 2016–2020 average. Almost fifty percent of this was designated for low-emission power generation, encompassing solar photovoltaic technology. Investment in fossil fuels climbed by 6%, totaling USD 33 billion in 2023, driven by huge demand for fuel and coal-fired electricity generation. Investment in clean energy is projected to double by 2030 under existing regulations, while an additional 20% increase is necessary to achieve the nation's energy and climate objectives. Reducing risks that increase the cost of capital will be essential in this endeavor (World Energy Investment 2024)

2.6 Conclusion

This paper examines India's current energy scenario. With population growth and rapid economic expansion, the country's energy demand is rising exponentially. Despite having a renewable energy potential of nearly 1001 GW, about 65% of current electricity needs are met by thermal energy, which relies on fossil fuels and increases environmental carbon levels. To reduce fossil fuel dependency, India purposes to produce 195 GW of renewable energy by 2025. Achieving this target requires annual growth rates of 50% in solar, 13% in wind, 2.7% in SHP, and 3.8% in biomass. The development rates for solar and wind energy are elevated, while those for small hydropower (SHP) and biomass are diminished, warranting focused attention on breakthroughs in solar and wind energy. The progression of renewable capacities encounters obstacles that must be resolved to achieve future objectives. These obstacles may be either universal or unique to renewable technologies. Enhanced legislative and regulatory frameworks, accessible financial assistance, and extensive awareness are crucial for advancing renewable energy. Key obstacles include a lack of advanced renewable technologies and the high cost of renewable products, necessitating increased R&D investment. Since most renewable products are imported, domestic manufacturing should be promoted to reduce costs. Additionally, the country faces a shortage of skilled personnel, requiring institutions to provide relevant training. Given the high cost of renewable products, the Indian government should offer monetary support with loans, loan rebates, and tariff reductions.

2.7 Limitation of the Study

The study examines the obstacles to renewable energy advancement in developing nations such as India and offers strategies to surmount these challenges. Nonetheless, these obstacles and suggestions have not been comprehensively implemented in other nations. Developed nations, characterized by greater renewable energy integration and sophisticated technologies, encounter less obstacles. In contrast, developing nations have greater obstacles. Moreover, nations with distinct climatic circumstances from India are unable to implement the identified obstacles and suggestions for renewable energy advancement.

2.8 Scope for Future Work

The obstacles and suggestions for advancing renewable energy in India may be appropriate to other comparable regions worldwide. This study can be reproduced in other nations to uniformly disseminate renewable energy and address environmental issues such as global warming. Specific impediments in India can be examined, and further solutions can be suggested to surmount these challenges.

Funding There is no funding for this research.

Data Availability Data available is not applicable.

Conflict of Interest No conflict of interest between authors.

References

Arto I (2016) The energy requirements of a developed world. Energy Sustain Dev 33:1–13. https://doi.org/10.1016/j.esd.2016.04.001

Avesh M, Hossain I, Sharma RC (2024) Revolutionizing transportation: the future impact of green energy. In: Dynamics of transportation ecosystem, modeling, and control. Energy, environment, and sustainability. Springer, Singapore, pp 261–293. https://doi.org/10.1007/978-981-97-0437-8_12

Bhattacharya SC, Jana C (2009) Renewable energy in India: historical developments and prospects. Energy 34(8):981–991. https://doi.org/10.1016/j.energy.2008.10.017

Blondeel M, Van de Graaf T (2018) Toward a global coal mining moratorium? A comparative analysis of coal mining policies in the USA, China, India and Australia. Clim Change 150(1):89–101. https://doi.org/10.1007/s10584-017-2135-5

BP Energy Outlook country and regional insights-India (2018). https://www.bp.com/content/dam/bp/en/corporate/pdf/energy-economics/energy-outlook/bp-energy-outlook-2018-country-insight-india.pdf.Accessed 30 Oct 2025

Kumar J CR., Majid MA (2020) Renewable energy for sustainable development in India: current status, future prospects, challenges, employment, and investment opportunities. Energy Sustain Soc 10(1). https://doi.org/10.1186/s13705-019-0232-1

DFG Analysis 2024–2025: power and new & renewable energy

Elavarasan RM, Shafiullah GM, Padmanaban S, Kumar NM, Annam A, Vetrichelvan AM, Holm-Nielsen JB et al (2020) A comprehensive review on renewable energy development, challenges, and policies of leading Indian states with an international perspective. IEEE Access 8:74432–74457

Flamos A (2010) The clean development mechanism—catalyst for wide spread deployment of renewable energy technologies? or misnomer? Environ Dev Sustain 12:89–102. https://doi.org/10.1007/s10668-008-9182-z

Hussain F, Dill H (2023) India incorporates green bonds into its climate finance strategy. World Bank Blogs. Available at https://blogs.worldbank.org/climatechange/india-incorporates-greenbonds-its-climate-finance-strategy. Last visited on 8–10–2023

IDFC LTD (2010) A discussion paper, barriers to development of renewable energy in India and proposed recommendations. Infrastructure Development Finance Company Ltd., February, Mumbai, India

IREDA (2023) Indian-Renewable-Energy-Development-Agency-Limited-Prospectus.pdf

Khare V, Nema S, Baredar P (2013) Status of solar wind renewable energy in India. Renew Sustain Energy Rev 27:1–10. https://doi.org/10.1016/j.rser.2013.06.018

Kumar JCR, Majid MA (2020) Renewable energy for sustainable development in India: current status, future prospects, challenges, employment, and investment opportunities. Energy, Sustain Society 10:2. https://doi.org/10.1186/s13705-019-0232-1

Kumar N, Pal N (2020) The existence of barriers and proposed recommendations for the development of renewable energy in Indian perspective. Environ Dev Sustain 22:2187–2205. https://doi.org/10.1007/s10668-018-0284-y

Kumar A, Kumar K, Kaushik N, Sharma S, Mishra S (2010) Renewable energy in India: current status and future potentials. Renew Sustain Energy Rev 14(8):2434–2442. https://doi.org/10.1016/j.rser.2010.04.003

Kumar NM, Chopra SS, Chand AA, Elavarasan RM, Shafiullah GM (2020) Hybrid renewable energy microgrid for a residential community: a techno-economic and environmental perspective in the context of the SDG7. Sustainability 12(10):3944

Kumar R, Goel R, Singh T, Mohanty SM, Gupta D, Alkhayyat A, Khanna R (2024a) Sustainable finance factors in Indian economy: analysis on policy of climate change and energy sector. Fluct Noise Lett 23(2)

Kumar V, Kumar R, Goel R, Singh T, Kaur MJ (2024b) The evolution of digital marketing technologies: a fuzzy TOPSIS approach to selected technology in E-commerce platform. Int J Syst Assurance Eng Manage:1–14

Ministry of Power (2023) https://powermin.gov.in/sites/default/files/National_Framework_for_promoting_Energy_Storage_Systems_August_2023.pdf

Mitigation CC (2011) IPCC special report on renewable energy sources and climate change mitigation. Renew Energy 20(11)

Mohanty SN, Singh T, Goel R, Baral SK, Kumar R (2024) A study on building awareness in cyber security for educational system in India using interpretive structural modellings. Int J Syst Assur Eng Manage:1–11

National Solar Water Pumping System (mnre.gov.in)

Nesamalar J, Venkatesh S, Raja S (2017) The drive of renewable energy in Tamilnadu: status, barriers and future prospect. Renew Sustain Energy Rev 73:115–124. https://doi.org/10.1016/j.rser.2017.01.123

Osorio LAR, Lobato MO, Castillo XAD (2005) Debates on sustainable development: towards a holistic view of reality. Environ Dev Sustain 7:501–518. https://doi.org/10.1007/s10668-004-5539-0

Press Information Bureau (2024) Cabinet approves PM-Surya Ghar: Muft Bijli Yojana for installing rooftop solar in one crore households. https://pib.gov.in/PressReleasePage.aspx?PRID=2010130

Thomas M (2015) India: early hydroelectric power—a piece of history. http://www.pipalpress.com/?P=1191. Accessed 26 Oct 2024

Uddin SN, Taplin R, Yu X (2010) Towards a sustainable energy future—exploring current barriers and potential solutions in Thailand. Environ Dev Sustain 12:63–87. https://doi.org/10.1007/s10668-008-9180-1

Union Ministry of Power (2022) Union Ministry of Power Annual Report 2022–23. Retrieved from https://cdnbbsr.s3waas.gov.in/s3716e1b8c6cd17b771da77391355749f3/uploads/2023/08/2023080211.pdf

World Energy Investment (2024). https://iea.blob.core.windows.net/assets/d829545d-fab6-4c98-b266-28556d86ce8d/WorldEnergyInvestment2024.pdf

World Energy (2013) Scenarios composing energy futures to 2050. World energy Council. https://www.worldenergy.org/wp-content/uploads/2013/09/World-Energy-Scenarios_Composing-energy-futures-to-2050_Full-report.pdf. Accessed 10 Oct 2024

World meters (2019). Available at http://www.worldometers.info/worldpopulation/india-population. Accessed 24 Oct 2024

Zeng S, Liu Y, Liu C, Nan X (2017) A review of renewable energy investment in the BRICS countries: history, models, problems and solutions. Renew Sustain Energy Rev 74:860–872. https://doi.org/10.1016/j.rser.2017.03.016

Chapter 3
Green Finance: A Path Towards Greener Future and Sustainable Development in Indian Banks

Vineet Chouhan and Ashish Porwal

Abstract The incorporation of green finance (GF) principles into sustainable development (SD) strategies has become a crucial approach to tackling urgent global issues in light of the growing environmental challenges. The study examines the intricate link between GF and SD, which also looks at how funding for environmentally friendly projects and activities advances more general developmental goals. This research offers a comprehensive literature review, case studies, and empirical data to clarify the opportunities, difficulties, and possible strategies for utilizing GF to hasten the transition to SD. The phrase "GF" was used to refer to SD consider both environmental and economic benefits. To enhance readers' comprehension of these essential factors, this study attempts to illustrate their theoretical aspects while providing a general overview of their fields and applications. Suggestions for making the process of SD easier to accomplish are also made, taking into account the necessity of green funding. Improved theoretical understanding of this field provided by this work can inspire academics to carry out varied study from many viewpoints and approaches, or, to put it another way, to provide new avenues for investigation.

Keywords GF · SD · Green financial products

3.1 Introduction

In today's world, SD that "aims to meet people's needs now without taking into account those of future generations" has gained significant attention. In order to harmonize ESG dynamics, the term was initially created over 25 years ago (IUCN 1980). It is impossible to build SD without making appropriate use of financial

V. Chouhan · A. Porwal (✉)
Institute of Law, Nirma University, Ahmedabad, India
e-mail: ashish.porwal@nirmauni.ac.in

V. Chouhan
e-mail: vineet.chouhan@nirmauni.ac.in

R. Kumar (ed.), *Sustainable Finance*, Sustainability Solutions,
https://doi.org/10.1007/978-3-031-97962-0_3

sources. The phrase "green finance" refers to a SD funding approach that is beneficial to the environment. Eco-friendly practices and the realms of finance and business come together in GF (Wang and Zhi 2016). In essence, GF is thought to be a key idea in SD since sustainability cannot be attained without ensuring appropriate and environmentally responsible use of financial resources. Environmental degradation is mostly caused by the residue of human activities such as transportation, manufacturing, agriculture, and energy generation. Greenhouse gas emissions and energy consumption have both increased dramatically over the last two decades, creating a major environmental crisis on a global scale. Energy is a basic necessity for both families and companies, and it is essential for any kind of economic activity to take place. Higher energy consumption will inevitably accompany rising globalization, which will increase carbon emissions (Shahbaz et al. 2018).

A loop of development is completed when the financial sector stimulates economic growth in an effort to entice FDI, which causes an increase in economic growth (Azam 2016). Finance costs are condensed when the financing network gets cavernous, which is encouraged by the progress of the finance sector. The Environmental Kuznets Curve (EKC) best captures this relationship among economic advancement and ecological integrity. Shocks to the financial sector affect entire economy, as the 2008–2009 financial crisis demonstrated. It is reasonable to argue that the financial industry is crucial to an economy's overall health. This begs the question, does financial development have a long-term effect on environmental degradation?

The financial sector's importance in SD will increase in the next years. GF is a principle that will promote continuous and sustainable economic growth. GF includes all financial instruments, such as green bonds, and financial entities, such as GB and GF, that only invest in ecologically sustainable activities and goods. Climate-smart finance, sometimes termed green financing, facilitates long-term carbon emission reductions and encourages the implementation of ecologically advantageous practices. The advancement of GF is propelled by market-oriented transformations (Mohanty et al. 2024). As per the World Economic Forum, investments in green infrastructure are expected to reach a total of $5 trillion per year by 2030, with the bulk of these investments coming from developing economies(Mir and Bhat, 2022). India must allocate more than $4.5 trillion on infrastructure by 2040 to support national renewable energy objectives, electric vehicles, and ecologically sustainable housing (Jha and Bakhshi, 2019).

3.2 Literature Reviews on Green Banking (GB) Services in India

GB refers to a kind of banking in which banks actively incorporate environmental sustainability into their daily operations, both internally and externally. Banks that prioritize this approach are sometimes referred to as green or sustainable banks (Afzal et al., 2022; Nosratabadi et al. 2020). Hossain, Rahman, Hossain, and Karim published a study in 2020. Sharma and Choubey (2022) emphasized that in the current competitive banking industry, it is essential for every bank to develop a new environmentally friendly product with increased stakeholder participation and sustainable development. Nevertheless, in their analysis of risks, Dharwal and Agarwal (2013) identified GB as a crucial strategy for mitigating many sorts of risks, including legal risk, credit risk, and reputation risk (Park and Kim 2020). In addition, they suggested other environmentally friendly behaviours, including engaging in carbon credit enterprises, investing in green financial products, opting for green mortgages, implementing measures for mitigating carbon emissions, enhancing energy awareness, implementing sustainable construction practices, and engaging in socially responsible initiatives. Khairunnessa et al. (2021) investigated the importance of sustaining financial viability in GB in Bangladesh, as mandated by the Bangladesh Bank (Khatun et al. 2021). In his study conducted in 2013, Meena found four advantages of green banking. These include the mitigation of deforestation, the promotion of environmental awareness among both employees and customers, the opportunity to profit from reduced interest rates, and the transformation of business practices to align with environmental goals. Shailaja and Olekar (2021) and Jaggi (2014) examined the GB initiatives and strategies of SBI and ICICI bank. SBI has launched many measures in this field, such as the Green Counter Channel, a heightened assurance to attaining carbon neutrality, digital remittance platforms, and the establishment of wind farms, among others. "ICICI's Green goods and services strategy" includes online banking for making it convenient and accessible for home and car finance. Furthermore, many financial institutions have adopted energy-saving methods, including double-sided-printing, recycling and the use of CFLs, among other tactics. Sharma and Choubey (2022) identified an absence of "GB services initiatives in India". The inspectors recommend that banks implement more sustainable financing methods and integrate economic and environmental considerations into their financing principles. This would require companies to engage in SD for societal benefit (Sheehy and Farneti 2021). Kalyar et al. (2019) performed an investigation on the impact of institutional pressures on financial institutions, finding that these institutions are adopting environmentally friendly regulations and practising transparent reporting in order to enhance their reputation in society (Kumar et al. 2024a). Institutional forces push managers to achieve environmental sustainability (Dai et al. 2021). In their study, Mendez and Houghton (2020) discovered that sustainable banking involves making investment and lending choices by considering the monitoring the environment and evaluating risks to satisfy sustainability requirements. Additionally, sustainable banking includes insurance services that specifically deal with environmental and climatic risks, which are important aspects of GF (Akomea-Frimpong et al. 2022). The study, Bukhari et al. (2020) examined the adoption of GB and its relationship with ESG factors. They specifically investigated how the interplay of factors influencing sustainability of environment. It resulted in the form of process that is affected by several environmental factors, and banks may influenced by executing firm actions in a systematic and uniform manner (Gunawan et al. 2022). Bukhari et al. (2020) assert that sustainable banking incorporates "GB methods to integrate ethical, social, and environmental considerations. Thus, as per Dec and Masiukiewicz (2021) to enhance the financial sector's sustainability, it is essential for the banking industry to give priority to provide finance for environmentally aware initiatives using financially viable banking methods. Gardi et al. (2020) augment the competitive edge of banks, elevate profits, optimize existing assets, and diminish costs related to invested capital and other expenditures. Until recently, GB was seen as a mere concept, and environmental considerations seemed to have little relevance to a bank's activities (Sharma and Choubey 2020). Historically, a bank's evaluation of a client's environmental practices was considered an overreach into private business. However, the prevailing sentiment now is that such assessments are essential to mitigate potential risks to the enterprise (Saxena et al. 2020). While financial organizations may not experience direct consequences from environmental deterioration, they nonetheless face indirect costs (Ding et al. 2022). If these banks do not embrace such steps, they will continue to be plagued by credit, legal, and reputation issues. The developing economies have not yet fully adopted the notion. Hafner et al. (2020) contend that research on GF in developing nations is significantly underdeveloped, necessitating a pressing need to fully explore and understand the notion. Similarly, Chen et al. (2022) expressed their worry about the significant research gap in green banking. Furthermore, Sharma and Choubey (2022) said that GB methods are not widely adopted in poor nations, including Indian banks. While GB has gained significant traction in developed countries, it has largely been overlooked in underdeveloped countries (Hermawan and Khoirunisa 2024; Rehman et al. 2021; Mejia-Escobar et al. 2020). There is generally a scarcity of research on green banking (GB) in the Indian context (Prakash et al. 2018). Empirical evidence suggests that Indian banks possess insufficient capacity to effectively implement environmentally sustainable banking practices (Bansal et al. 2023; Nethala et al. 2022). The RBI has a crucial responsibility in promoting and enhancing environmental standards. India, as a developing country, should prioritize the social dimension of banking and establish a connection between it and economic progress (Khalid et al. 2021). The main related study topics in India include CSR, environmental management and sustainable banking strategies in private and public sector financial institutions. Multiple researchers have conducted thorough investigations on these subjects. (Bansal et al. 2023; Nethala et al. 2022; Khalid et al. 2021). There exists a notable disparity between the objectives that banks want to endorse and the public's perception of their actions in relation to GB (Park and Kim 2020; Sharma 2022). Consequently, GB literature in India is lacking. Since it was the pioneer in giving GB projects priority, the SBI was selected for this examination. The majority of GB studies have either examined actual GB procedures or examined how customers and bankers see GB (Rahman et al. 2023; Rehman et al. 2021; Niekerk, 2024). The stated problem needs to be further investigated due to this disparity. As a result, our study intends to fill a knowledge vacuum in this field.

The SBI (largest commercial bank in India) paved the way for the "green banking" movement and spearheaded stricter environmental standards. It was SBI that pioneered the Coimbatore wind farm project (Kumar et al. 2024b). The concept of GB has pushed banks to become more tech-savvy and paperless in their offerings, all the while reducing their environmental footprint and fulfilling their corporate social responsibility to support national development. Comprehending the need for environmentally friendly initiatives is essential as it will decide the success or failure of these investments depending on the level of satisfaction shown by environmentally conscious clients. Additionally, they assist banks in fostering a reputation for environmental accountability, a quality that is becoming progressively crucial in the contemporary world. GB programs usually involve ride-sharing, recycling, energy efficiency, and ecologically conscious loans (Wang et al. 2022). Owing to strict environmental laws that are enforced by qualified authorities worldwide, for industries to operate, several requirements would need to be met. It enhances authorities' and customers' mental capacities to reflect environmental sensibility. GB increases the nation's GDP while reducing expenses and using less energy based upon this literature this research paper analyses the Nexus Between AI and GF for SD in Indian Banks.

3.3 Methodology

It was appropriate for the study objectives and the distinct characteristics of the population examined, which pertains to the Indian banking industry. With the aim of presenting empirical data from the Indian banking industry, on the relationship between GF and SD, we used a structured questionnaire as a research tool or instrument.

3.4 Demographical Profile

For the data collection form bank managers, we successfully distributed survey questions to 3460 bank managers across the country by dividing the sample geographical area into five reasons, northern, southern, eastern, western, and central side of the country. 1260 questionnaires were completed in response to the study and out of it completely filled in total 990 are included in the sample size. However, we discarded 270 surveys due to inadequate or inappropriate replies. The study's final sample consisted of 990 valid responses from bank managers in the banking industry. To ensure the representativeness of sample, we utilized Cronbach's alpha (α) as an accepted reliability measure and with the value of $0.876 > 0.700$ for this research and support its findings. Demography of the individuals is shown in Table 3.1.

The demographic profile of respondents encompassed variables such as gender, age, educational qualification,and professional experience. Notably, the surveys were administered to participants drawn from a diverse range of backgrounds, thereby ensuring representation across different socio-economic and professional strata (see Table 3.1). It includes the 66.6% of male bank managers, with the highest young and dynamic managers with 38%, professional qualification holder managers with the 48.8% and a good experience one with 5–10 years of experience of 39% respondents. Therefore, the results obtained from this sample provide a realistic representation of

Table 3.1 Respondent's demography

Category	Detail	Freq.	Per cent
Gender	Male	659	66.6
	Female	331	33.4
Age	20–30	276	27.9
	30–40	376	38
	40–50	223	22.5
	50–60	115	11.6
Highest qualification	Graduate	108	10.9
	Postgraduate	223	22.5
	Professional	483	48.8
	Doctorate	176	17.8
Experiences (in years)	0–5	283	28.6
	05–10	386	39
	10–15	190	19.2
	15–20	131	13.2

the diverse viewpoints within the sample, as influenced by their demographic traits. Furthermore, the survey was conducted in five regions of India, involving participants from a society characterized by diverse cultural and behavioural characteristics. From this viewpoint, the obtained sample is sufficiently diverse to create a picture that accurately represents a wide range of traits.

3.5 Research Tool

This study used a questionnaire including 24 closed-ended questions with multiple-choice response possibilities, deemed adequate for the specified research aims. Additionally, we constructed the 24 questions in a clear and direct way. The inquiries were controlled among bank managers at banking sector in India from February to June 2023. The questionnaire includes six questions from environmental practices, six from the social practices, and four from economic practices with four questions from the green finances practices, and finally, four questions were included in the portion of improvement in green future. The questions inquired about scaling or ranking, allowing respondents to provide ordinal or ranked values to the given responses using a Likert scale (5-point). Moreover, the use of questionnaires has sparked debates both in support and opposition, mostly concerning methodological factors. Table 3.2 presents the scale items used in the study as under:

Table 3.2 Scale items and its sources

Item	Definition	Source
EcV1	Enhancement of competitive advantage	Nenavath (2022), Falcone (2020), Tu and Wu (2021)
EcV2	Reducing investment and other expenses Increasing the generation of revenue	Lee (2020), Tu et al. (2021)
EcV3	Enhanced financial contribution to the government treasury	Ozili (2022), Lee (2020)
EcV4	Generation of additional economic value in the form of economic value added	Kaur et al. (2020), Laukkanen and Tura (2020)
EcV5	Enhancement of current assets with the addition of capital	Li et al. (2022), Hu et al. (2021)
EcV6	Decrease in overall risk (banking as well as security risk)	Akomea-Frimpong et al. (2022), Khan et al. (2022), Li et al. (2022)
SV1	Improvement of the bank's reputation	Akomea-Frimpong et al. (2022), Chen et al. (2022)
SV2	Formation of trust	Gilchrist et al.(2021)
SV3	Plans for engaging stakeholders	Mohanty et al. (2023), Tran et al. (2020)
SV4	Enhancement of brand recognition	Cai and Song (2024), Zhang et al. (2022a, b)
SV5	Enhancement of consumer delight	Cai and Song (2024), Zhang et al. (2022a, b)
SV6	Enhancement of employee benefits	YinYin Lau et al. (2023), Agubata et al. (2022),Zheng et al. (2021)
EV1	Reduction in energy demand for banking services	Du (2023), Li et al. (2021)
EV2	Reduction in external energy usage	Ning et al. (2023), Liu et al. (2022)
EV3	Decrease in carbon emissions resulting from banking operations	Cai and Song (2024), Tran et al. (2020)
EV4	Convenient availability of funding via green bonds/ loans	Ning et al. (2023), Zhang et al. (2022a, b)
GFI1	Green bonds	Liu et al. (2022), Zhang et al. (2022a, b), Tran et al. (2020)
GFI2	Green loans	Zhang et al. (2022a, b), Tran et al. (2020)
GFI3	Sustainable Investment Funds or ESG Funds	Zhang et al. (2022a, b), Liu et al. (2022), Tran et al. (2020)
GFI4	Green insurance	Xiong and Sun (2023), Akomea-Frimpong et al. (2022), Debrah et al. (2022)

(continued)

Table 3.2 (continued)

Item	Definition	Source
IGF1	Implementing GF has a substantial positive impact on bank's revenue and market share and finally it enhances the reputation	Chen et al. (2022), Gilchrist et al. (2021), Zheng et al. (2021)
IGF2	GF substantially reduces the operational costs of bank by on reducing both paper waste and energy consumption within our bank	Cai et al. (2024), Gilchrist et al. (2021)
IGF3	GF enhances the connection between the community and stakeholders	Gilchrist et al. (2021), Chygryn et al. (2020)
IGF4	GF enhances banks' adherence to environmental regulations	Xiong and Sun (2023), Tran et al. (2020)

3.6 Data Analysis Method

SEM is a statistical instrument used to identify relation among variables within a multivariate context. The method employs are SEM to ascertain the structural link between observable and latent variables. The first survey data was analysed using SPSS with AMOS for comprehensive statistical evaluation. SEM was employed to validate the hypotheses proposed. Furthermore, we conducted both EFA and CFA to evaluate the precision, reliability, computation, and structural models(Hair et al. 2010). The standardized root mean residual (SRMR) quantifies the results by calculating the square root of the variance between the correlation matrices of the sample and the model. The model is deemed a good match if the SRMR is less than 0.08.

3.7 Findings

3.7.1 EFA Outcomes

Table 3.3 gives the "descriptives, factor extraction loadings with reliabilities" for the variables employed in the research. It presents the results of the EFA and CFA, using factor analysis. Cattell (1966) used EFA on a set of 24 questions to assess the construct's factor loadings. The determination of the optimal factors was based on the application of eigenvalue criteria. The reliability and validity of the data were evaluated using α coefficients. The KMO test, with a value of 0.918, exceeded the minimum acceptable threshold of 0.5, suggesting that sample size is adequate. Bartlett's test of sphericity indicated the correlation matrix's significance ($p < 0.000$), as evidenced by a chi-square value of 3103.819. Furthermore, the combined variance explained 68.59% of the dimension including green financing and the attributes of sustainability performance. The communalities were found above the standard of 0.500. The reliability of the constructs was assessed using α coefficients, which yielded values ranging from 0.712 to 0.876, showing acceptance levels of internal consistency and measurement precision.

3.7.2 Discriminant Validity

It was confirmed by comparing the square root of the AVE with its corresponding inter-factor correlations. From Table 3.4, it can be noted that the AVE for each construct was more than the squared inter-factor correlation for that construct, confirming discriminant validity of the constructs. The three-dimensional model of GF is demonstrated with acceptable model fit as well as robust reliability, discriminant, and convergent validity.

3.7.3 Measurement Model

After conducting the investigation, Gerbing and Anderson (1988) confirmed the measurement model by using the data from the CFA. Subsequently, the model was evaluated using standardized coefficients and other model fit indicators. Table 3.2 displays the findings, and how well the GF factors work and how well they do for sustainability was assessed using conventional estimations of factor loadings. In the CFA, the convergent validity of the GF dimensions was assessed using Cronbach's α coefficients, which ranged from 0.583 to 0.881, thereby exceeding the commonly accepted threshold value of 0.50. The study evaluated the sustainability performance and reliability of the GF elements in Indian banking by employing α, composite reliability (CR), and AVE. The α coefficients for the ESG performance dimensions surpassed the lowest acceptable edge of 0.70. However, the α coefficient for the ENV dimension was slightly lower at 0.692, as observed in Table 3.4.

Figure 3.1 displays the CFA measurement model's results used in the investigation, including the standardized estimates. Therefore, the measures' convergent validity is deemed satisfactory and adequate. Given that the AVE values were > 0.50 and the CR values exceeded 0.60, the constructs were considered to be both valid and reliable. Table 3.5 presents the GoF indices, which indicate acceptable model fit. The model fit statistics were as follows: χ^2/df = 2.513; GFI = 0.912, RMSEA = 0.039; CFI = 0.9565; NFI = 0.859, AGFI = 0.891; and p-value = 0.000. While the NFI value of 0.899 did not reach the criteria of 0.90, they are still considered acceptable since they are close to the reference value. Bentler and Bonett (1980) states that the GFI, CFI, NFI, and AGFI should have a minimum value of 0.90 to be regarded acceptable. Additionally, an RMSEA value of $\leq$ 0.05 is treated as suitable. The results suggest that 24 variables from the first-order CFA model of GF analysis of the aspects and banking sustainability performance in India which was suitable as study sample. Hence, the entire model's adequacy and satisfaction are confirmed.

Table 3.3 EFA outcomes and descriptive

Variables	Items	Mean	Standard deviation	Factor loadings					Eigen value	Cronbach's Alpha (α)	CR	AVE
				EcV	SC	EV	GFI	IGF				
Economic dimension (EcV)	EcV1	3.91	1.002	0.827					3.187	0.837	0.843	0.437
	EcV2	3.82	0.986	0.802								
	EcV3	3.8	0.875	0.777								
	EcV4	3.98	0.817	0.728								
	EcV5	3.78	0.841	0.705								
	EcV6	3.87	0.997	0.698								
Social dimension (SC)	SOC1	4.21	0.84		0.587				3.025	0.876	0.881	0.558
	SOC2	4.2	0.88		0.742							
	SOC3	3.89	0.799		0.819							
	SOC4	4.07	0.892		0.703							
	SOC5	4.18	0.921		0.674							
	SOC6	3.91	0.914		0.572							
Environmental dimension (EV)	EV1	3.89	0.875			0.636			1.576	0.712	0.693	0.431
	EV2	3.61	0.87			0.599						
	EV3	3.78	0.954			0.686						
	EV4	3.93	0.929			0.665						
GF indicators (GFI)	GFI1	3.99	0.852				0.576		2.813	0.863	0.864	0.517
	GFI2	4.19	0.816				0.559					
	GFI3	4.09	0.868				0.644					
	GFI4	3.99	0.917				0.622					
Improvement in green future (IGF)	IGF1	4.02	-0.02					0.641	1.452	0.869		

(continued)

Table 3.3 (continued)

Variables	Items	Mean	Standard deviation	Factor loadings					Eigen value	Cronbach's Alpha (α)	CR	AVE
				EcV	SC	EV	GFI	IGF				
	IGF2	3.76	0.879					0.615			0.882	0.557
	IGF3	4.01	0.865					0.566				
	IGF4	3.29	0.791					0.532				

Principal component analysis (PCA) was utilized for data reduction. The KMO measure of sampling adequacy was calculated at 0.926, confirming the dataset's suitability for factor analysis. Bartlett's test of sphericity yielded a significant result ($p < 0.001$); it supported factor analysis. The extracted factors accounted for 68.59% of the total variance.

Source Authors (2024)

Table 3.4 Validity measurement

Dimension	Mean	SD	ECO	SOC	ENV	GFI	IGF
Economic (EcV)	3.84	0.65	0.641				
Social (SC)	4.07	0.69	0.473	0.727			
Environmental (EV)	3.77	0.74	0.386	0.418	0.686		
GF indicators (GFI)	4	0.66	0.376	0.371	0.484	0.699	
Improvement in green future (IGF)	4.65	0.62	0.373	0.352	0.542	0.722	0.755

Source Test results, Authors (2024)

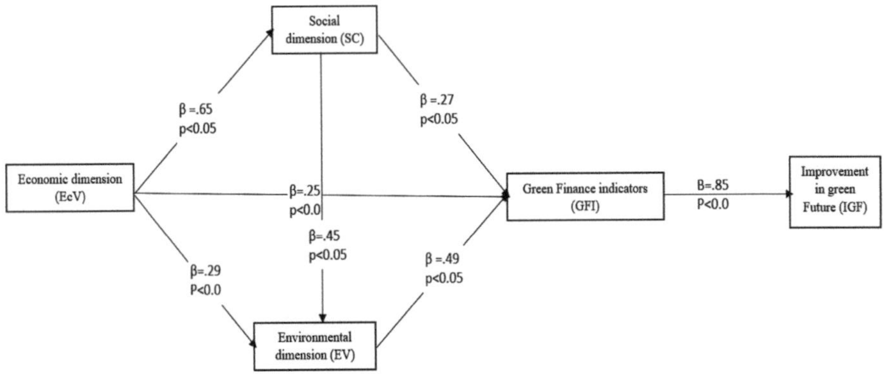

Fig. 3.1 SEM model for greener future and SD

Table 3.5 GoF statistics for models

Variable	χ^2/df	GFI	RMSEA	CFI	NFI	AGFI	*p*
Model for measurement	2.513	0.912	0.039	0.956	0.899	0.91	0
Structure model	2.515	0.912	0.039	0.957	0.891	0.911	0

Both models, each comprising 24 items, were evaluated using standard goodness-of-fit indices. The following cut-off criteria were employed: χ^2/df < 3; GFI > 0.9; CFI > 0.90; NFI > 0.90; AGFI > 0.90 and RMSEA < 0.08. The results indicated model fit's acceptability with χ^2/df = 2.513, GFI = 0.912; RMSEA = 0.039, CFI = 0.9565, NFI = 0.859, and AGFI = 0.891.

3.7.4 SEM Model

Figure 3.1 illustrates the SEM, with demonstration of influence for relationship among the internal components and ideas. The results showed the various sections of GF having a beneficial and substantial effect on SB within the Indian banking

Table 3.6 Hypothesis test result

Impact	Beta	*z*-Test	*p*	Result of relation
H1: Eco. → Soc	0.65***	6.66	0.000	Confirmed
H2: Eco. → Env	0.29***	2.752	0.000	Confirmed
H3: Eco. → Gfi	0.25***	1.43	0.006	Confirmed
H4: Soc. → Gfi	0.27***	1.067	0.001	Confirmed
H5: Soc. → Env	0.45***	3.01	0.007	Confirmed
H6: Env. → Gfi	0.49***	2.941	0.002	Confirmed
H7: Gfi. → Igf	0.85***	2.941	0.000	Confirmed

Notes ***$p < 0.05$; SPSS AMOS outcome by Authors' 2024

industry. As a result, the EcV and SC dimension positively affect the GFI ($\beta = 0.65$, $p < 0.05$) and ENV ($\beta = 0.29$, $p < 0.00$). In addition, the EcV has statistically significant positive influence on GFI ($\beta = 0.25$, $p > 0.05$). Similarly, the SC dimension also exert a statistically significant impact on GFI ($\beta = 0.27$, $p < 0.05$), along with EV ($\beta = 0.49$, $p < 0.05$). Likewise, influence of social dimension (SC) on environmental (EV) is favourable and has a substantial statistical effect ($\beta = 0.45$, $p < 0.05$). The GFI exert a statistically significant beneficial impact ($\beta = 0.85$, $p < 0.05$) on the improvement in green future (IGF). In summary, the empirical results demonstrate that the three aspects of GF significantly and positively impact the GF Indicators (GFI). Similarly, the GFI also have a positive influence on the improvement in green future (IGF), which contributes to the sustainability of the banking in India. Table 3.6 displays the conclusive results, representing a significantly positive association among the three extents of GFI and both IGF and the performance for sustainability of the SCBs in India.

To assess the significance of GF for the banking growth in India, participants were queried with the following inquiries that "GF is an essential element of your banks' growth strategy?". The findings revealed that 92.78% of the bankers replied positively, whilst just 7.12% held the belief that GF does not contribute meaningfully to the long-term growth of their institutions in a responsible manner. Thus, it may be inferred that GFI is indispensable for the improved green future for developing banking sector in India.

3.8 Future of GF in India

Given that sustainability of environment is a major global concern, there is more opportunity to invest in green initiatives that use renewable energy sources. As a result, a lot of banks and other financial organizations would be interested in entering this expanding market. As a result, demand for structured green funds and green bonds will rise. Moreover, purchasing such bonds would provide investors with the benefit of diversification. For the purpose of adapting to and mitigating change of climate, a PM's Council on Climate Change is developed at 2007, which underwent restructuring in 2014. This council has spearheaded various initiatives, including the National Clean Energy Fund, the Jawaharlal Nehru National Mission for Solar, for Enhanced Energy Efficiency, National Action Plan for Climate Change and the on Strategic Knowledge for Climate Change. Additional initiatives include Expert Groups on Low Carbon Strategies, Auto Fuel Vision and Policy 2025, etc. In 2015, NABARD was recognized as a National Implementing Entity (NIE) by the Green Climate Fund (GCF), a financial mechanism under the UNFCCC for climate mitigation and adaptation projects in India.

Indian aims to envision 450 GW of "renewable energy capacity" by 2030, and this requires a speedy transition to sources, such as solar, wind, and hydropower for clean energy and that would result in large investments in infrastructure and renewable energy projects, opening up new prospects for green finance. India's high rates of economic expansion are outpacing the country's energy supply. Frequent power outages and ongoing power shortages are the outcome of this. The Indian government is placing more emphasis on our efforts to mitigate climate change and promotion of renewable sources of energy as well as increasing measures of energy efficiency, focusing on MSME sectors; they contribute a lot in the economy of India. MSMEs frequently overlook energy efficiency in particular because they have limited access to suitable finance instruments and technical know-how.

3.9 Conclusion and Suggestions

This research seeks to understand how corporate governance mechanisms influence bank sustainability performance in India. It provides an overview of green financing trends from 2021 to 2024 and analyses the impact of specific GF characteristics on bank sustainability performance, such as climate risk management, primary data was collected from bankers employed in India. The empirical findings demonstrate the credibility of the sustainable dimensions in relation to GF indicators, as well as the beneficial influence of GF indicators on the improvement of a sustainable future.

Based on these results, it is evident that three aspects of green future, as "ESE dimensions", are interconnected with the concepts of Environmental, Social, Governance (ESG) standards and SDGs. These connections may be further expanded to enhance the progress towards a greener future for Indian banks. These results suggest three components of GF have important real-world consequences for Indian banks. To strengthen their sustainability performance, they must focus on investing in eco-friendly initiatives. The research also found a significant positive correlation between all three specified banks parameters in India. Our study suggests that GF has beneficial effects on several stakeholders, such as bankers, managers, suppliers, academics, and communities. It facilitates the attainment of environmental sustainability by using tactics such as energy preservation, digital banking, electronic transactions, staff involvement initiatives, and training and growth opportunities.

To assess the importance of GF for the growth of the Indian banking sector, participants were queried about whether they consider green financing to be a crucial element of their banks' development plan. The findings revealed that 92.78% of the bankers provided a positive response, whilst just 7.12% held the belief that GF does not significantly affect the SD of their institutions. Further, the model explains that GFI represents the 85% of the impact over the GF adoption in Indian banks. Thus, it can be inferred that GF is crucial for building sustainable and resilient future, characterized by economic growth with advancement of the banking industry in India.

To ensure sustainable economic growth, India needs to develop green financing products and methods. By 2040, India will require over $4.5 trillion in infrastructure funding to develop its economy sustainably. India has already started investing in GF goods and services with support from the public, commercial, and government sectors. The amount of money we now have invested in GF may not be sufficient to meet the SD objective. The Indian government ought to establish an open and honest green investment plan with a focus on long-term, comprehensive economic perspectives. Furthermore, for GF rules and regulations to draw in investors, they must be transparent and unambiguous. The implementation of a GF idea must draw in both domestic and foreign investors. GB is therefore the way and basis for SD, also advantageous to banks, businesses, and the environment in general. Therefore, by removing the obstacles and raising corporate citizens' understanding climate change challenges and resource scarcity, India has a significant opportunity to invest in green infrastructure and mobilize GF. We hope that GF will become widely accepted in all societies worldwide in the near future.

Funding NA.

Conflict of Interest There is no conflict of interest.

References

Afzal A, Rasoulinezhad E, Malik Z (2022) Green finance and sustainable development in Europe. Econ Res-Ekonomska Istraživanja 35(1):5150–5163

Agubata NS, Okolo LI, Ogwu SO (2022) Effect of employee benefits on organizational growth of consumer goods firms in Nigeria. Studia Universitatis Vasile Goldiş Arad, Seria Ştiinţe Economice 32(3):116–137

Akomea-Frimpong I, Adeabah D, Ofosu D, Tenakwah EJ (2022) A review of studies on green finance of banks, research gaps and future directions. J Sustain Fin Invest 12(4):1241–1264

Azam M (2016) Does environmental degradation shackle economic growth? A panel data investigation on 11 Asian countries. Renew Sustain Energy Rev 65:175–182

Bansal N, Taneja S, Ozen E (2023) Green financing as a bridge between green banking strategies and environmental performance in Punjab, India. Int J Sustain Dev Plan 18(10):35–42

Bentler PM, Bonett DG (1980) Significance tests and goodness of fit in the analysis of covariance structures. Psychol Bull 88:588–606

Bukhari SAA, Hashim F, Amran A (2020) Green banking: a road map for adoption. Int J Ethics Syst 36(3):371–385
Cai S (2024) Research on the impact of green finance on ecological industrial structure. Acad J Bus Manag 6(2):243–247
Cai X, Song X (2024) Towards sustainable environment: unleashing the mechanism between green finance and corporate social responsibility. Energy Environ 35(2):986–1003
Cattell RB (1966) The scree test for the number of factors. Multivar Behav Res 1(2):245–276. https://doi.org/10.1207/s15327906mbr0102_10
Chen J, Siddik AB, Zheng GW, Masukujjaman M, Bekhzod S (2022) The effect of green banking practices on banks' environmental performance and green financing: an empirical study. Energies 15(4):1292
Chygryn O, Bilan Y, Kwilinski A (2020) Stakeholders of green competitiveness: innovative approaches for creating communicative system. Mark Manage Innov 3:358–370
Dai J, Xie L, Chu Z (2021) Developing sustainable supply chain management: the interplay of institutional pressures and sustainability capabilities. Sustain Prod Consump 28:254–268
Debrah C, Chan APC, Darko A (2022). Green finance gap in green buildings: a scoping review and future research needs. Build Environ 207(A):108443, 1–19
Dec P, Masiukiewicz P (2021) Socially responsible financial products as a contribution of financial institutions to sustainable development. Sustainability 13(6):3067
Dharwal M, Agarwal A (2013) Green banking: an innovative initiative for sustainable development. ACCMAN Inst Manag Art 2(3):1–7
Ding X, Appolloni A, Shahzad M (2022) Environmental administrative penalty, corporate environmental disclosures and the cost of debt. J Clean Prod 332:129919
Du G (2023) Nexus between green finance, renewable energy, and carbon intensity in selected Asian countries. J Clean Prod 405:136822
Falcone PM (2020) Environmental regulation and green investments: the role of green finance. Int J Green Econ 14(2):159–173
Gardi B, Hamawandy NM, Vian Sulaiman Hama Saeed RMA, Sulaiman AA, Mahmood SA, Al-Kake FA (2020) The effect of capital competence on the profitability of development and investment banks in Turkey. Solid State Technol 63(6):12571–12583
Gerbing DW, Anderson JC (1988) An updated paradigm for scale development incorporating unidimensionality and its assessment. J Mark Res 25(2):186–192
Gilchrist D, Yu J, Zhong R (2021) The limits of green finance: a survey of literature in the context of green bonds and green loans. Sustainability 13(2), 478:1–12
Gunawan J, Permatasari P, Sharma U (2022) Exploring sustainability and green banking disclosures: a study of banking sector. Environ Dev Sustain 24(9):11153–11194
Hafner S, Jones A, Anger-Kraavi A, Pohl J (2020) Closing the green finance gap–a systems perspective. Environ Innov Soc Trans 34:26–60
Hair JF, Black WC, Babin BJ, Anderson RE (2010) Multivariate data analysis, 4th edn. Prentice Hall, Hoboken, NJ, USA
Hermawan S, Khoirunisa ZA (2024) The emergence of green banking: a sustainable financing strategy for protecting against deforestation in ASEAN. J Environ Dev 33(1):96–124
Hu J, Li J, Li X, Liu Y, Wang W, Zheng L (2021) Will green finance contribute to a green recovery? Evidence from green financial pilot zone in China. Front Public Health 9:794195
Jaggi G (2014) Green banking: Initiatives by SBI and ICICI. Paripex-Indian J Res 3(6):2691–1361
Jha B, Bakhshi P (2019) Green finance: fostering sustainable development in India. Int J Recent Technol Eng 8(4):3798–3801
Kalyar MN, Shoukat A, Shafique I (2019) Enhancing firms' environmental performance and financial performance through green supply chain management practices and institutional pressures. Sustain Acc Manage Policy J 11(2):451–476
Kaur J, Sarma AK, Jha MK, Gera P (2020) Valorisation of crude glycerol to value-added products: perspectives of process technology, economics and environmental issues. Biotechnol Reports 27:e00487

Khairunnessa F, Vazquez-Brust DA, Yakovleva N (2021) A review of the recent developments of green banking in Bangladesh. Sustainability 13(4):1904

Khalid AM, Sharma S, Dubey AK (2021) Concerns of developing countries and the sustainable development goals: case for India. Int J Sust Dev World 28(4):303–315

Khan MA, Riaz H, Ahmed M, Saeed A (2022) Does green finance really deliver what is expected? An empirical perspective. Borsa Istanbul Rev 22(3):586–593

Khatun MN, Sarker MNI, Mitra S (2021) Green banking and sustainable development in Bangladesh. Sustain Clim Change 14(5):262–271

Kumar R, Goel R, Singh T, Mohanty SM, Gupta D, Alkhayyat A, Khanna R (2024a) Sustainable finance factors in Indian economy: analysis on policy of climate change and energy sector. Fluct Noise Lett 23(2)

Kumar V, Kumar R, Goel R, Singh T, Kaur MJ (2024b) The evolution of digital marketing technologies: a fuzzy TOPSIS approach to selected technology in E-commerce platform. Int J Syst Assurance Eng Manage:1–14

Laukkanen M, Tura N (2020) The potential of sharing economy business models for sustainable value creation. J Clean Prod 253:120004

Lee JW (2020) Green finance and sustainable development goals: the case of China. J Asian Fin Econ Bus 7(7):577–586

Li M, Hamawandy NM, Wahid F, Rjoub H, Bao Z (2021) Renewable energy resources investment and green finance: evidence from China. Resour Policy 74:102402

Li Z, Kuo TH, Siao-Yun W, Vinh LT (2022) Role of green finance, volatility and risk in promoting the investments in renewable energy resources in the post-covid-19. Resour Policy 76:102563

Liu H, Yao P, Latif S, Aslam S, Iqbal N (2022) Impact of Green financing, FinTech, and financial inclusion on energy efficiency. Environ Sci Poll Res:1–12

Mejia-Escobar JC, González-Ruiz JD, Duque-Grisales E (2020) Sustainable financial products in the Latin America banking industry: current status and insights. Sustainability 12(14):5648

Mendez A, Houghton DP (2020) Sustainable banking: the role of multilateral development banks as norm entrepreneurs. Sustainability 12(3):972

Mir AA, Bhat AA (2022) Green banking and sustainability–a review. Arab Gulf J Sci Res 40(3):247–263

Mohanty S, Nanda SS, Soubhari T, Biswal S, Patnaik S (2023) Emerging research trends in green finance: a bibliometric overview. J Risk Fin Manage 16(2):108

Mohanty SN, Singh T, Goel R, Baral SK, Kumar R (2024) A study on building awareness in cyber security for educational system in India using interpretive structural modellings. Int J Syst Assurance Eng Manage:1–11

Nenavath S (2022) Impact of fintech and green finance on environmental quality protection in India: by applying the semi-parametric difference-in-differences (SDID). Renew Energy 193:913–919

Nethala VJ, Pathan MFI, Sekhar MSC (2022) A study on cooperative banks in India with special reference to marketing strategies. J Contemp Issues Bus Govern 28(4):584–593

Ning Y, Cherian J, Sial MS, Álvarez-Otero S, Comite U, Zia-Ud-Din M (2023) Green bond as a new determinant of sustainable green financing, energy efficiency investment, and economic growth: a global perspective. Environ Sci Poll Res 30(22):61324–61339

Nosratabadi S, Pinter G, Mosavi A, Semperger S (2020) Sustainable banking; evaluation of the European business models. Sustainability 12(6):2314

Ozili PK (2022) Green finance research around the world: a review of literature. Int J Green Econ 16(1):56–75

Park H, Kim JD (2020) Transition towards green banking: role of financial regulators and financial institutions. Asian J Sustain Social Responsibility 5(1):1–25

Prakash A, Kumar K, Srivastava A (2018) Consolidation in the Indian banking sector: evaluation of sustainable development readiness of the public sector banks in India. Int J Sustain Strateg Manag 6(1):3–16

Rahman MH, Rahman J, Tanchangya T, Esquivias MA (2023) Green banking initiatives and sustainability: a comparative analysis between Bangladesh and India. Res Globalization:100184

Rehman A, Ullah I, Afridi FEA, Ullah Z, Zeeshan M, Hussain A, Rahman HU (2021) Adoption of green banking practices and environmental performance in Pakistan: a demonstration of structural equation modelling. Environ Dev Sustain:1–21

Saxena N, Hayes E, Bertino E, Ojo P, Choo KKR, Burnap P (2020) Impact and key challenges of insider threats on organizations and critical businesses. Electronics 9(9):1460

Shahbaz M, Nasir MA, Roubaud D (2018) Environmental degradation in France: the effects of FDI, financial development, and energy innovations. Energy Econ 74:843–857

Shailaja D, Olekar RO (2021) Green banking practices in India as an environmental perspective-with reference to SBI and ICICI. IJRAR-Int J Res Anal Rev (IJRAR) 8(1):91–100

Sharma SE (2022) The international political economy of green finance: a research agenda. J Environ Policy Plan. Advance online publication

Sharma M, Choubey A (2020) Resistance towards green banking adoption in Delhi NCR: a grounded theory approach. AIMA J Manage Res 14(2/4):1–20

Sharma M, Choubey A (2022) Green banking initiatives: a qualitative study on Indian banking sector. Environ Dev Sustain 24(1):293–319

Sheehy B, Farneti F (2021) Corporate social responsibility, sustainability, sustainable development and corporate sustainability: what is the difference, and does it matter? Sustainability 13(11):5965

Tran T, Do H, Vu T, Do N (2020) The factors affecting green investment for sustainable development. Decis Sci Lett 9(3):365–386

Tu Y, Wu W (2021) How does green innovation improve enterprises' competitive advantage? The role of organizational learning. Sustain Prod Consump 26:504–516

Tu Q, Mo J, Liu Z, Gong C, Fan Y (2021) Using green finance to counteract the adverse effects of COVID-19 pandemic on renewable energy investment-the case of offshore wind power in China. Energy Policy 158:112542

van Niekerk AJ (2024) Economic inclusion: green finance and the SDGs. Sustainability 16(3):1128

Wang Y, Zhi Q (2016) The role of green finance in environmental protection: two aspects of market mechanism and policies. Energy Procedia 104:311–316

Wang KH, Zhao YX, Jiang CF, Li ZZ (2022) Does green finance inspire sustainable development? Evidence from a global perspective. Econ Anal Policy 75:412–426

World Wildlife Fund (1980) World conservation strategy: living resource conservation for sustainable development, vol 1. IUCN, Gland, Switzerland

Xiong Q, Sun D (2023) Influence analysis of green finance development impact on carbon emissions: an exploratory study based on fsQCA. Environ Sci Pollut Res Int 30(22):61369–61380

YinYin Lau P, Park S, Hsu YC, Lien BYH, Ho JA (2023) Does investment in green employee development climate matter for environmental commitment and green well-being? A case study of a palm oil company in Malaysia. SAGE Open 13(4):21582440231204130

Zhang W, Liu X, Liu J, Zhou Y (2022a) Endogenous development of green finance and cultivation mechanism of green bankers. Environ Sci Poll Res:1–11

Zhang X, Wang Z, Zhong X, Yang S, Siddik AB (2022b) Do green banking activities improve the banks' environmental performance? The mediating effect of green financing. Sustainability 14(2):989

Zheng GW, Siddik AB, Masukujjaman M, Fatema N (2021) Factors affecting the sustainability performance of financial institutions in Bangladesh: the role of green finance. Sustainability 13(18):10165

Chapter 4
Fuzzy ISM Approach to Assessment of Factors Affecting to Adoption Industry 5.0 Technologies for Mitigation of Climate Risk

Rakesh Kumar and Bijay Kumar Paikaray

Abstract Aim of this study is to focus on important factor that supports to adoption of Industry 5.0 for reducing climate risk. Industry 5.0, a technology-driven strategy, has raised concerns about social issues like resource waste and worker unemployment. The circular economy, which recycles waste back into production, is a key focus in sustainable development. Industry 5.0 offers technical solutions to support the circular economy, improving supply chain sustainability, transparency, waste management, and resource efficiency. This study selected ten critical base on previous literature review. Fuzzy ISM is used for data analysis. A strategic roadmap identifies ten major Industry 5.0 drivers, emphasising resilience, sustainability, and people orientation, to support climate risk reduction. The roadmap aims to create a sustainable supply chain network and fortify the enterprise ecosystem.

Keywords Industry 5.0 · Climate risk · Sustainable finance · Smart technology · Sustainability

4.1 Introduction

Given that traditional energy sources contribute to both economic shortfalls and environmental degradation, energy sustainability is a critical component of sustainable development (SD). Fossil fuels can be replaced by cleaner, more dependable energy supplies from renewable sources. A more sustainable future may result from the judicious use of renewable energy sources. Proponents of Industry 5.0 contend that Industry 4.0 is insufficient because it ignores the profit-driven basis of

R. Kumar (✉)
Department of Management Studies, Graphic Era (Deemed to be) University, Dehradun, India
e-mail: aigrakeshkumar@gmail.com

B. K. Paikaray
Department of Computer Science and Engineering, Centre for Data Science, Siksha 'O' Anusandhan (Deemed to be) University, Bhubaneswar, Odisha, India

R. Kumar (ed.), *Sustainable Finance*, Sustainability Solutions,
https://doi.org/10.1007/978-3-031-97962-0_4

contemporary production and consumer economic structures in favour of concentrating on technology-driven productivity and production efficiency. The goal of Industry 5.0 is to establish a data-driven, hyperconnected industrial environment that prioritises sustainable development. Industry 5.0 (I5.0) and blockchain technology (BT) are related, fostering traceability, transparency, and decentralisation. The decentralised decision-making and improved human–robot collaboration of Industry 5.0 are compatible with BT's distributed ledger technology (Durán et al. 2024). Blockchain smart contracts interface with Industry 5.0's automation concepts to provide safe and automated contracts. Both technologies' tamper-resistant features address trust and security in human–machine cooperation. A disruptive synergy aiming at accomplishing sustainable supply chain objectives is represented by the integration of BT and I5.0 in the Bangladeshi RMG sector. While Industry 5.0's collaborative approach encourages accountability, empowers workers, and improves human–machine collaboration, blockchain's distributed ledger guarantees transparency and traceability (Al Amin and Baldacci 2024). Industry 4.0 and Industry 5.0 are developing quickly, and the IoT's explosive development and the volume of data generated present scalability and security issues (Wang et al. 2024). By facilitating safe, legitimate, and dispersed IoT networks and encouraging automation, dependability, and openness of information, technology can support the implementation of CE. Additionally, it may increase supply chain circular performance, gather waste data for reverse logistics operations, speed up the promotion of CE, and strengthen auditing capabilities for production and purchase decisions. In order to create a more sustainable society, blockchain is also essential for the implementation of the metaverse and extended reality. Blockchain technology has excellent uses in both eras (Kannan et al. 2024). The revolutionary technology known as Industry 5.0 is predicted to result in more responsible work, employment losses, and a rise in digital skills. Concerns have been raised regarding the effects on digital culture, employment, and legal and normative frameworks. The transition is not well understood by many specialists, which breeds scepticism and possible hazards. These difficulties may worsen if upper management does not assist them. Governments should implement laws to encourage the adoption of Industry 4.0, and a strong legal framework is essential for handling cybersecurity and data privacy concerns. In an Industry 4.0 setting, collaboration is especially crucial since it promotes unity and boosts output (Fernández et al. 2024).

4.1.1 Motive of the Research

The environment has been greatly damaged by human civilisation and industrialisation, which has resulted in increased energy use and rising urbanisation. Numerous environmental problems, such as pollution, dangers, and ecosystem deterioration, have been brought on by this. Protecting the environment, natural resources, and future generations requires sustainability, which entails combining short- and long-term issues (Ghobakhloo et al. 2024). Ecological fairness and preventing resource depletion are the goals of sustainable development. Urbanisation and globalisation speed up consumption and pollution, which causes climate change, making global warming a serious problem (Awasthi et al. 2023). To lower greenhouse gas emissions and provide a sustainable environment, decarbonisation is required (Chandra and Shukla 2024). Prior research has concentrated on sustainable energy, green supply chain management, and individual sustainability aspects; this study intends to address additional sustainability enablers through Industry 5.0 that reduced climate risk (Ghobakhloo et al. 2022). The objectives of this study are as follows:-

1. To determined important factored affecting to adoption of Industry 5.0 technologies for reduction of climate risk.
2. To prioritise criteria to adoption Industry 5.0 technologies for reduction of climate change risk.

4.2 Literature Review

Industry 5.0 is a socio-technical phenomenon that focuses on digitalising industrial value networks and utilising technologies like cloud computing, big data analytics, and AI. It supports autonomy, equality, privacy, and human dignity. Enterprise resource planning (ERP) is a crucial enabling technology, providing real-time insights into supply chain and corporate processes. However, the model relies on disruptive technical advancements to achieve its goals (Hsu et al. 2024a, b, c). The possibility of fire threats is further increased by the employment environment, which includes overcrowding, insufficient fire safety measures, and inadequate ventilation. Many factories, nonetheless, are still beyond the purview of the Accord and are unable to meet all fire safety requirements. Fire safety regulations can be put into place by clothing manufacturers to lower incidences and guarantee worker welfare, both of which are essential for sustainability. By preventing property damage, lowering potential financial losses, lowering insurance costs, and fostering environmental sustainability, putting fire safety procedures into place can assist economic sustainability. Long-term advantages, pressure from both domestic and foreign sources, and cultural shifts are all contributing to the slow development of the literature on fire safety analysis in the clothing industry. A more scientific and methodical approach is required to analyse fire safety concerns in the clothing industry (Siraj et al. 2023).

The goal of Industry 5.0 is to overcome the shortcomings of the current innovation environment by upgrading Industry 4.0 in a gradual yet evolutionary manner. It is distinguished by human-centred industrial processes that are powered by innovative technology that enhance employment, productivity, and working conditions. Early research recognised Industry 4.0's drawbacks, including its technological centricity and digital divide. In 2021, the European Commission unveiled its Industry 5.0 agenda, which expanded the framework to give new socio-environmental demands first priority. Early in 2022, however, the European Commission contended that Industry 4.0 was not the best foundation for dealing with social unrest and ecological concerns. A new vision for the industry is represented by Industry 5.0, which redefines the purpose and role of business models, value chains, and digital transformation in the highly interconnected corporate environment (Bai et al. 2023).

By fusing the intelligence of humans and machines, Industry 5.0, a group of new technologies that emerged from Industry 4.0, seeks to improve process efficiency. Artificial intelligence plays a significant role now a days in air quality and prediction (Awasthi et al. 2024), which align with Industry 4.0 principle to boost systems and upgrade environmental outcomes. This entails returning human workers to factory floors, integrating intelligent systems with workflows, and continuously observing human–machine interaction. Cloud-based energy solutions, nanotechnology, artificial intelligence, and quantum computing define Industry 6.0, a more futuristic framework. It includes highly customised class one products, dynamic supply chain management, mass customisation, and hyper-connectivity between industries (Raut et al. 2018). Industry 5.0 seeks to reduce waste and lessen its effects on the environment by establishing a circular economy and increasing resource efficiency. Technological change must be sustainable and in line with the objectives of socioeconomic growth. Sustainability is aided by renewable resources, IoT, and AI. For operational effectiveness and environmental stewardship, CDG focuses on integrating cutting-edge digital technologies (Bagherian et al. 2024) (Table 4.1).

Table 4.1 Selection of criteria

Criteria	Description	References
Accessibility (C1)	This refers to easy use of smart technology device. This is a challenge to shift from traditional structure to new set up	Caiado et al. (2024)
Overall performance (C2)	The adoption of smart technology device depends on performance in organisation like operational performance, economic output, etc.	Engelbrecht et al. (2024)
Initial cost (C3)	Affordable cost of equipment is an important factor to adoption of smart device in circular economy process	Bai et al. (2023)
Sustainability (C4)	The smart device shall more sustainable and environment-friendly	Hsu et al. (2024a, b, c)
Data management (C5)	It is an important requirement to track and managing of data in organisation	Zekhnini et al. (2024)
Smart system (C6)	Smart system device reduces time and operational cost	Jamil et al. (2024)
Customer loyalty (C7)	Customer loyalty enhances adaptability of smart device	Hsu et al. (2024a, b, c)
Health and safety of people (C8)	Environment-friendly and healthier equipment helps in sustainability	Nazir et al. (2024)
Government regulation (C9)	Positive government support helps to adoption of sustainable smart technology	Jamil et al. (2024)
Productivity (C10)	Sustainable smart technology shall enhance overall productivity in organisation	Jamil et al. (2024)

4.3 Methodology

This study finds out important factor which affect investment of sustainable finance in Industry 5.0 technologies for reducing of climate risk. Smart technology is an important facet to reduce climate risk. The study emphasise how these technologies can adoption in business organisation. The study selected 15 experts from industry and academic. This study included 10 different factors based on previous literature review and experts opinions. The FUZYY ISM techniques are used for prioritisation of criteria. When adopting Industry 5.0 to shift to a circular economy, businesses need to take their resource constraints into account. Achieving industry commonality requires quantifying norms and indicators. Supply chain monitoring and high-level support enhance the synergy between technology and the circular economy. Support from government policies is necessary for sustainable development. Industry 5.0 skills require multi-function specialisation and employee training. Innovation and environmental preservation are promoted when enterprise governance adopts a circular economy.

4.4 Result

Product life cycles are extended and sustainable business models are supported by using Industry 5.0 technologies and the circular economy to create recyclable goods. Supply chain partner analysis increases transparency, maximises non-value-added links, and builds up expertise with circular production. Tables 4.2, 4.3, 4.4, 4.5 and 4.6 show result of fuzzy ISM. Table 4.2 shows structural self- nitration matric (SSIM).

After SSIM, Table 4.3 indicates reachability metric (RM). This helps to determine driving power of variable.

Table 4.2 SSIM

Variables	1	2	3	4	5	6	7	8	9	10
Accessibility (C1)		V	V	V	V	V	V	V	V	V
Overall performance (C2)			V	V	V	V	V	V	V	V
Initial cost (C3)				V	V	V	V	V	V	V
Sustainability (C4)					V	V	V	V	V	V
Data management (C5)						A	V	A	V	V
Smart system (C6)							V	V	V	V
Customer loyalty (C7)								V	V	V
Health and safety of people (C8)									V	V
Government regulation (C9)										A
Productivity (C10)										

Table 4.3 Reachability matrix (RM)

Variables	1	2	3	4	5	6	7	8	9	10	Driving power
Accessibility	1	1	1	1	1	1	1	1	1	1	10
Overall performance	0	1	1	1	1	1	1	1	1	1	9
Initial cost	0	0	1	1	1	1	1	1	1	1	8
Sustainability	0	0	0	1	1	1	1	1	1	1	7
Data management	0	0	0	0	1	0	1	0	1	1	4
Smart system	0	0	0	0	1	1	1	1	1	1	6
Customer loyalty	0	0	0	0	0	0	1	1	1	1	4
Health and safety of people	0	0	0	0	1	0	0	1	1	1	4
Government regulation	0	0	0	0	0	0	0	0	1	0	1
Productivity	0	0	0	0	0	0	0	0	1	1	2
Dependence power	1	2	3	4	7	5	7	7	10	9	

Table 4.4 Final reachability matrix (FRM)

Variables	1	2	3	4	5	6	7	8	9	10	Driving power
Accessibility (C1)	1	1	1	1	1	1	1	1	1	1	10
Overall performance (C2)	0	1	1	1	1	1	1	1	1	1	9
Initial cost (C3)	0	0	1	1	1	1	1	1	1	1	8
Sustainability (C4)	0	0	0	1	1	1	1	1	1	1	7
Data management (C5)	0	0	0	0	1	0	1	1*	1	1	5
Smart system (C6)	0	0	0	0	1	1	1	1	1	1	6
Customer loyalty (C7)	0	0	0	0	1*	0	1	1	1	1	5
Health and safety of people (C8)	0	0	0	0	1	0	1*	1	1	1	5
Government regulation (C9)	0	0	0	0	0	0	0	0	1	0	1
Productivity (C10)	0	0	0	0	0	0	0	0	1	1	2
Dependence power	1	2	3	4	8	5	8	8	10	9	

Asterisk provides the transitive relation between factors in final reachability matrix

Table 4.5 Level partitioning (LP)

Elements (Mi)	Reachability set *R*(Mi)	Antecedent set *A*(Ni)	Intersection set *R*(Mi) ∩ *A*(Ni)	Level
1	1	1	1	8
2	2	1, 2	2	7
3	3	1, 2, 3	3	6
4	4	1, 2, 3, 4	4	5
5	5, 7, 8	1, 2, 3, 4, 5, 6, 7, 8	5, 7, 8	3
6	6	1, 2, 3, 4, 6	6	4
7	5, 7, 8	1, 2, 3, 4, 5, 6, 7, 8	5, 7, 8	3
8	5, 7, 8	1, 2, 3, 4, 5, 6, 7, 8	5, 7, 8	3
9	9	1, 2, 3, 4, 5, 6, 7, 8, 9, 10	9	1
10	10	1, 2, 3, 4, 5, 6, 7, 8, 10	10	2

Table 4.6 Level partitioning iterations

Elements (Mi)	MI	NI	Intersection	Level
1	1, 2, 3, 4, 5, 6, 7, 8, 9, 10	1	1	
2	2, 3, 4, 5, 6, 7, 8, 9, 10	1, 2	2	
3	3, 4, 5, 6, 7, 8, 9, 10	1, 2, 3	3	
4	4, 5, 6, 7, 8, 9, 10	1, 2, 3, 4	4	
5	5, 7, 8, 9, 10	1, 2, 3, 4, 5, 6, 7, 8	5, 7, 8	
6	5, 6, 7, 8, 9, 10	1, 2, 3, 4, 6	6	
7	5, 7, 8, 9, 10	1, 2, 3, 4, 5, 6, 7, 8	5, 7, 8	
8	5, 7, 8, 9, 10	1, 2, 3, 4, 5, 6, 7, 8	5, 7, 8	
9	9,	1, 2, 3, 4, 5, 6, 7, 8, 9, 10	9	1
10	9, 10	1, 2, 3, 4, 5, 6, 7, 8, 10	10	

After final reachability matrix, level partitioning is next step. Table 4.5 indicates level petitioning.

4.5 Discussion

The study focus of important driver of ESG which support to adoption of Industry 5.0 for climate risk reduction. Fuzzy ISM is used for determination of most important criteria. 8 levels determined in the study are shown in Fig. 4.1. Level 1 is highest level, i.e. government regulation. Productivity at level 2. Data management, customer loyalty, and health and safety of people at level 3. Smart systems, sustainability, initial

Table 4.7 Conical matrix (CM)

Variables	9	10	5	7	8	6	4	3	2	1	Driving power	Level
9	1	0	0	0	0	0	0	0	0	0	1	1
10	1	1	0	0	0	0	0	0	0	0	2	2
5	1	1	1	1	1*	0	0	0	0	0	5	3
7	1	1	1*	1	1	0	0	0	0	0	5	3
8	1	1	1	1*	1	0	0	0	0	0	5	3
6	1	1	1	1	1	1	0	0	0	0	6	4
4	1	1	1	1	1	1	1	0	0	0	7	5
3	1	1	1	1	1	1	1	1	0	0	8	6
2	1	1	1	1	1	1	1	1	1	0	9	7
1	1	1	1	1	1	1	1	1	1	1	10	8
Dependence power	10	9	8	8	8	5	4	3	2	1		
Level	1	2	3	3	3	4	5	6	7	8		

Asterisk provides the transitive relation between factors in final reachability matrix

coat, overall performance, and accessibility are levels 4, 5, 6, 7, and 8, respectively. Research on Industry 5.0 is still in its early stages; thus, further studies are required to confirm its influence on reducing climate risk. Future developments in science and technology may render the existing drivers obsolete. Future studies should examine the connections between variables and offer a more effective strategic road map to entrepreneurs and legislators. It is imperative that company strategy incorporates the circular economy concept under Industry 5.0. However, for pragmatic reasons and because businesses and society must be ready, putting the roadmap into action may prove difficult (Table 4.8).

4.6 Finding and Recommendation

The technology-driven strategy known as "Industry 4.0" has sparked worries about social problems like resource waste and worker unemployment. In contrast, Industry 5.0 tackles problems like resource scarcity and environmental degradation with an emphasis on human sustainability and resilience. It places a strong emphasis on employee wellbeing, environmental preservation, and social welfare. One of the most important topics in sustainable development is the circular economy, which emphasises recycling trash back into the production process. By improving supply chain sustainability, transparency, waste management, and resource efficiency, Industry 5.0 provides technical solutions to assist the circular economy. This integration helps businesses achieve both environmental sustainability and economic growth, and it speeds up the green shift.

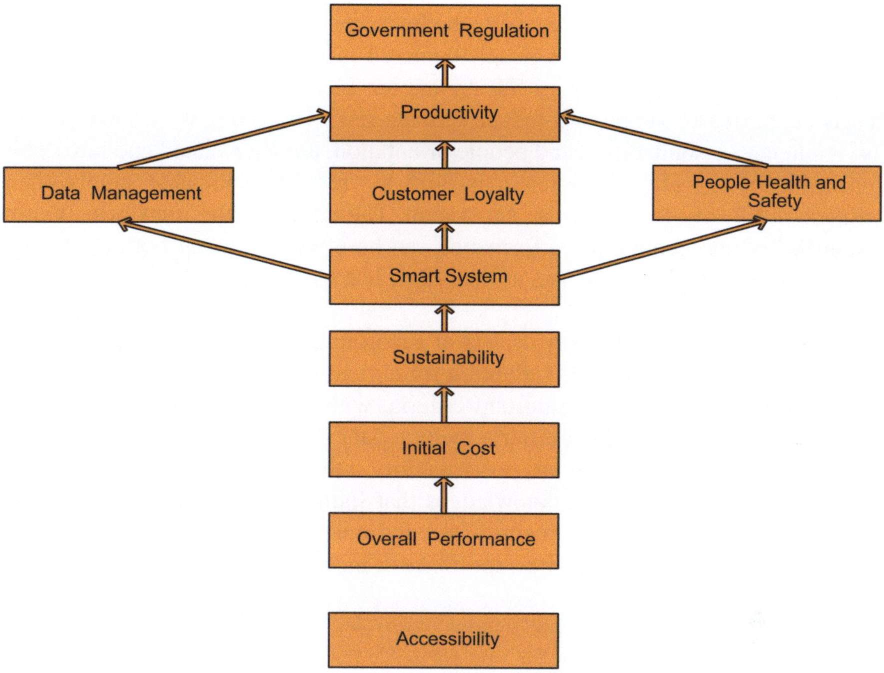

Fig. 4.1 Level of the criteria

Table 4.8 Reduced conical matrix (CM)

Variables	9	10	5	7	8	6	4	3	2	1	Driving power	Level
Government regulation (C9)	1	0	0	0	0	0	0	0	0	0	1	1
Productivity (C10)	1	1	0	0	0	0	0	0	0	0	2	2
Data management (C5)	0	1	1	1	1*	0	0	0	0	0	5	3
Customer loyalty (C7)	0	1	1*	1	1	0	0	0	0	0	5	3
Health and safety of people (C8)	0	1	1	1*	1	0	0	0	0	0	5	3
Smart system (C6)	0	0	1	1	1	1	0	0	0	0	6	4
Sustainability (C4)	0	0	0	0	0	1	1	0	0	0	7	5
Initial cost (C3)	0	0	0	0	0	0	1	1	0	0	8	6
Overall performance(C2)	0	0	0	0	0	0	0	1	1	0	9	7
Accessibility (C1)	0	0	0	0	0	0	0	0	1	1	10	8
Dependence power	10	9	8	8	8	5	4	3	2	1		
Level	1	2	3	3	3	4	5	6	7	8		

A circular economy, which converts the linear model into a closed-loop one, and Industry 5.0, which emphasises people-cantered models, are essential for sustainable growth. Ten major Industry 5.0 drivers are identified in a strategic roadmap that is suggested to support to reduce climate risk. These drivers, which emphasise resilience, sustainability, and people orientation, are interrelated and have a big impact on the transition. The roadmap aims to create a sustainable supply chain network and fortify the enterprise ecosystem. The effects of the transformation on sustainability and economic performance can be evaluated using both qualitative and quantitative methods, such as an analytical hierarchy approach for digital green technologies.

The study focuses on the key drivers of Environmental, Social, and Governance (ESG) that support Industry 5.0 adoption for climate risk reduction. It uses a fuzzy ISM to determine the most important criteria, with government regulation at the highest level. Other important criteria include data management, customer loyalty, health and safety, smart systems, sustainability, initial coat, overall performance, and accessibility. The study acknowledges that Industry 5.0 research is still in its early stages, and future developments in science and technology may render existing drivers obsolete. It emphasises the importance of incorporating the circular economy concept in company strategies.

4.7 Conclusion

Industry 4.0, a technology-driven strategy, has raised concerns about social issues like resource waste and worker unemployment. In contrast, Industry 5.0 focuses on human sustainability and resilience, emphasising employee wellbeing, environmental preservation, and social welfare. The circular economy, which recycles waste back into production, is a key focus in sustainable development. Industry 5.0 offers technical solutions to support the circular economy, improving supply chain sustainability, transparency, waste management, and resource efficiency. A strategic roadmap identifies ten major Industry 5.0 drivers, focusing on resilience, sustainability, and people orientation, to support climate risk reduction. The roadmap aims to create a sustainable supply chain network and fortify the enterprise ecosystem. The study focuses on key ESG drivers supporting Industry 5.0 adoption for climate risk reduction, using fuzzy ISM criteria, including government regulation, data management, customer loyalty, health and safety, smart systems, sustainability, initial coat, overall performance, and accessibility.

Funding NA.

Conflict of Interest There is no conflict of interest for this research.

References

Al Amin M, Baldacci R (2024) Blockchain technology and Industry 5.0 synergy for sustainable development in RMG industries: an ISM and fuzzy DEMATEL approach. Discov Sustain 5(1):1–26

Awasthi A, Pattnayak KC, Tandon A, Sarkar A, Chakraborty M (2023) Implications of climate change on surface temperature in North Indian states: evidence from CMIP6 model ensembles. Front Environ Sci 11:1264757

Awasthi A, Pattnayak KC, Tiwari PR, Panda SK, Gautam S, Choudhury T, Sar A (2024) Air quality monitoring (AQM) and prediction: transitioning from conventional to AI techniques. In: Artificial intelligence for air quality monitoring and prediction. CRC Press, pp 1–23

Bagherian A, Gershon M, Kumar S, Mishra MK (2024) Analyzing the relationship between digitalization and energy sustainability: a comprehensive ISM-MICMAC and DEMATEL approach. Expert Syst Appl 236:121193

Bai C, Zhou H, Sarkis J (2023) Evaluating Industry 4.0 technology and sustainable development goals–a social perspective. Int J Prod Res 61(23):8094–8114

Caiado RGG, Machado E, Santos RS, Thomé AMT, Scavarda LF (2024) Sustainable I4. 0 integration and transition to I5.0 in traditional and digital technological organisations. Technol Forecast Social Change 207:123582

Chandra A, Shukla R (2024) ISM-fuzzy MICMAC approach for modelling the enablers of sustainability. IIMBG J Sustain Bus Innov 2(2):120–142

Durán C, Yazdi AK, Derpich I, Tan Y (2024) Leveraging blockchain for maritime port supply chain management through multicriteria decision making. Mathematics 12(10):1511

Engelbrecht A, Yitmen I, Almusaed A (2024) Industry 4.0 transformation towards Industry 5.0 paradigm-challenges, opportunities and practices

Fernández I, Puente J, Ponte B, Gómez A (2024) Integration of AHP and fuzzy inference systems for empowering transformative journeys in organizations: assessing the implementation of Industry 4.0 in SMEs. Appl Intell:1–21

Ghobakhloo M, Iranmanesh M, Mubarak MF, Mubarik M, Rejeb A, Nilashi M (2022) Identifying industry 5.0 contributions to sustainable development: a strategy roadmap for delivering sustainability values. Sustain Prod Consump 33:716–737

Ghobakhloo M, Iranmanesh M, Foroughi B, Rejeb A, Nikbin D, Tseng ML (2024) A practical guide on strategic roadmapping for information and operations technology management: a case study on industry 5.0 transformation. J Industr Prod Eng 41(5):397–421

Hsu CH, Li ZH, Zhuo HJ, Zhang TY (2024) Enabling Industry 5.0-driven circular economy transformation: a strategic roadmap. Sustainability 16(22):9954

Hsu CH, Liu JC, Cai XQ, Zhang TY, Lv WY (2024) Enabling sustainable diffusion in supply chains through Industry 5.0: an impact analysis of key enablers for SMEs in emerging economies. Mathematics 12(24):3938

Hsu CH, Wu JZ, Zhang TY, Chen JY (2024c) Deploying Industry 5.0 drivers to enhance sustainable supply chain risk resilience. Int J Sustain Eng 17(1):1–28

Jamil MA, Mustofa R, Hossain NUI, Rahman SA, Chowdhury S (2024) A structural equation modeling framework for exploring the industry 5.0 and sustainable supply chain determinants. Supply Chain Anal 6:100060

Kannan D, Amiri AS, Shaayesteh MT, Nasr AK, Mina H (2024) Unveiling barriers to the integration of blockchain-based circular economy and Industry 5.0 in manufacturing industries: a strategic prioritization approach. Bus Strat Environ

Nazir H, Kamal KA, Abbas A, Raza M, Yahya M, Ibrahim M (2024) Unlocking insights: structural equation modeling approach to unveil the nexus between Industry 5.0 evolution and sustainable supply chain dynamics. Kurdish Stud 12(4):1095–1107

Raut RD, Priyadarshinee P, Gardas BB, Jha MK (2018) Analyzing the factors influencing cloud computing adoption using three stage hybrid SEM-ANN-ISM (SEANIS) approach. Technol Forecast Soc Chang 134:98–123

Siraj MT, Debnath B, Payel SB, Bari AM, Islam ARMT (2023) Analysis of the fire risks and mitigation approaches in the apparel manufacturing industry: implications toward operational safety and sustainability. Heliyon 9(9)

Wang W, Wang Y, Chen Y, Deveci M, Kadry S, Pedrycz W (2024) Analyzing the barriers to resilience supply chain adoption in the food industry using hybrid interval-valued Fermatean fuzzy PROMETHEE-II model. J Ind Inf Integr 40:100614

Zekhnini K, Chaouni Benabdellah A, Bag S, Gupta S (2024) Supply chain 5.0 digitalization: an integrated approach for risk assessment. Manage Decis

Chapter 5
Sustainable Investment Policies in Renewable Energy Through Artificial Intelligence

Pankaj Kumar and S. K. Singh

Abstract It is more crucial than ever to make the switch to a low-carbon/sustainable/green economy. The growing renewable energy industries, which demands large investments, are at the heart of this transformation. Because artificial intelligence (AI) can analyze large volume of data, find different patterns, and optimize sophisticated systems, it is emerging as an ergastic tool to support sustainable investment policies in the region globally. Harnessing the power of AI can help the increase decision-making clarity, reduce risks, and accelerate the implementation of renewable energy projects for investors and governments. This chapter explores the synergistic nature of AI and renewable energy, demonstrating how this technology may be applied to appreciate project viability, optimize investment strategies, and develop efficient energy policies. The challenges to sustainable investment policies in renewable energy include data quality and availability, costs associated with those investments, regulatory issues, inherent AI challenges, ethical issues, and issues in scaling and applying AI-based predictive models. Last but not least, implementing AI into sustainable investments may foster a new potential for greater, sustainable and environment friendly world.

Keywords AI · Investment policies · Sustainable investments · Sustainable energies · Renewable energies

5.1 Introduction

This is the art and science of investing money so that we get financial returns without compromising on social or environmental impact; it is called sustainable investing. The same is true for how much of a mess the world has become, as people grow more

P. Kumar (✉)
Department of Computer Science and Engineering, G. L. Bajaj Institute of Technology and Management, Greater Noida, Uttar Pradesh, India
e-mail: unpankaj@gmail.com

S. K. Singh
Sanskaram University, Jhajjar, Haryana, India

R. Kumar (ed.), *Sustainable Finance*, Sustainability Solutions,
https://doi.org/10.1007/978-3-031-97962-0_5

aware that we live in lines with environmental crisis and resources shortage; environmentally conscious investments are becoming quite efficient. Within this larger framework, renewable energy (solar, wind, hydro, and geothermal) has continued to rise in importance for sustainable investment.

5.1.1 The Need for Renewable Energy

The world's energy sector is one of the main sources of greenhouse gas emissions leading to climate change. Fossil fuels like coal, oil, and gas are not only devastatingly detrimental to the environment; they will also run out sooner or later. Therefore, renewable energy sources offer an alternative that can produce a possible abundant and clean way of getting more energy as practicing local demand (Danish and Senjyu 2023; Ferdaus et al. 2024). Therefore, in order to reduce both carbon emissions and global warming for the long term of our planet, governments have made significant investments on renewable energy. Following are the different types of renewable energy:

5.1.1.1 Solar Energy

- **Photovoltaic (PV) Solar Power**: It uses solar panels to convert sunlight directly into electricity.
- **Solar Thermal Energy**: It uses sunlight to heat a fluid, which then produces steam to generate electricity.

5.1.1.2 Wind Energy

- **Onshore Wind**: The turbines located on land that convert wind into electricity.
- **Offshore Wind**: The wind turbines placed in bodies of water, typically on the continental shelf, where wind speeds are generally higher.

5.1.1.3 Hydropower

- **Large Hydropower**: The dams and reservoirs store water, which is released to generate electricity.
- **Small Hydropower**: The smaller scale projects that divert water flow through a turbine.
- **Run-of-the-River**: It uses the natural flow of rivers without significant storage, with minimal impact on the environment.

5.1.1.4 Geothermal Energy

- **Direct Use Applications**: Using geothermal heat directly for heating buildings, greenhouses, etc.
- **Geothermal Power Plants**: Extracting heat from the earth to produce steam that drives turbines to generate electricity.

5.1.1.5 Biomass Energy

- **Bioenergy**: The energy produced from organic materials such as wood, agricultural residues, and waste.
- **Biofuels**: The liquid fuels like ethanol and biodiesel produced from organic materials, used primarily for transportation.

5.1.1.6 Ocean Energy

- **Tidal Energy**: Harvesting energy from the movement of tides.
- **Wave Energy**: Capturing energy from surface waves.
- **Ocean Thermal Energy Conversion (OTEC)**: Uses the temperature difference between warmer surface water and colder deep water to generate electricity.

5.1.1.7 Hydrogen Energy (from Renewable Sources)

- **Green Hydrogen**: Hydrogen produced using renewable energy sources like wind or solar to electrolyze water, splitting it into hydrogen and oxygen.

5.1.2 The Role of Sustainable Investment

As alluded earlier, the renewable sources always need support from sustainable investment to replace traditional fossil fuel energy with greener alternatives, which means money to build, install, and grow green tech (Kou et al. 2023; Selim 2020). By allocating capital to these sectors, investors can contribute toward ensuring they grow in line with global sustainability objectives—such as the Paris Agreement's goal of limiting warming to well below 2 °C.

5.1.3 Types of Sustainable Investments in Renewable Energy

Aside from this, investments made in the renewable energy sector can take different forms, such as:

- **Direct Investments**: Directly making investments in renewable energy sources such as wind farms, solar power generation systems, and hydropower plants.
- **Green Bonds**: Buying green bonds that are assigned specific purposes, such as financing green projects or the renewable energy sector.
- **Sustainable Funds**: Investing in fuel cell systems, growing solar, and putting funds in exchange-traded funds (ETFs), or mutual funds, or other financial structured products that promote the business with superior environmental, social and governance (ESG) aspects.
- **Venture Capital and Private Equity**: Providing assistance to nascent enterprises and burgeoning firms that are engaged in the creation of novel renewable energy technologies.

5.1.4 Drivers of Sustainable Investment in Renewable Energy

Several factors are driving the growth of sustainable investment in renewable energy:

- **Policy Support**: Globally, governments are progressively enacting policies, like tax breaks, requirements for renewable energy, and subsidies, to stimulate investment in clean energy.
- **Technological Advancements**: Innovations in renewable energy technologies are making them more efficient, cost-effective, and scalable, attracting more investors.
- **Consumer Demand**: While people become more concerned with environmental problems and deterioration environmentally friendly electricity, as well as for corporate organizations, are on the rise.
- **Risk Management**: Now, they know there is severe financial risk related to climate change, and they see acquisition of renewable power as a way to mitigate risk.

5.1.5 Investment Policies for Renewable Energy

Pervasive big data processing, massive computational power, information technology, and refined ML/DL algorithms are some of the causes for the recent advancements in AI technologies (Kumar and Singh 2020). The combination of renewable energy with another technological area, AI presents a great opportunity that can fuel sustainable investment. Policy-makers and investors may reduce risks, advance the transition to a zero-carbon economy, and make more accurate decisions by applying AI in its data processing, estimation, and control. In this chapter, the different ways in which artificial intelligence or AI could enhance sustainable investment in renewable energy policies are explored (Kumar and Singh 2017).

AI has an enormous potential in changing the way sustainable investments are being made and especially in renewable energy sector (Hannan et al. 2021). If the power of the EU is used to its full potential, much more satisfactory, successful and effective policies to pursue the change can be made. Here is some ways AI might improve investment policies for renewable energy:

5.1.5.1 Risk Assessment and Mitigation

- **Predictive Analytics**: AI is capable of analyzing huge amount of data and identifying potential threats, for example, storms, hardware crash, or shift in the market tendencies.
- **Portfolio Optimization**: By defining high and low risk investments, it means that use of AI can help in portfolio diversification with the aim of decreasing the levels of risk involved.
- **Insurance Pricing**: Machine learning models in AI are capable of evaluating the risk of renewable energy projects hence positive impacts toward premium insurance rates.

5.1.5.2 Policy Development and Evaluation

- **Policy impact assessment**: To this end, AI and network analysis can enable evaluation of outcomes of various policy options and their effects on investment in renewable energy sources, integration of the renewable energy sources into the utility grid and efficiency of the entire energy system.
- **Regulatory optimization**: AI can identify regulatory barriers and recommend policy changes to accelerate renewable energy deployment.
- **Incentive design**: AI can help design effective incentive programs to stimulate investment in renewable energy technologies.

5.1.5.3 Market Analysis and Forecasting

- **Demand Forecasting**: Besides, it can also predict trends of future energy demand which will help investors to make better decisions on the type, size and areas where renewable energy is needed.
- **Price Forecasting**: AI can forecast energy prices, enabling investors to hedge against market volatility.
- **Technology Assessment**: AI can track the development of renewable energy technologies and identify emerging opportunities.

5.1.6 AI Tools and Technologies

Through the description above, it can be deduced that AI technology has a very vital role in increasing several elements of investing in renewable energy in the following ways. Here are a few particular AI tools and technologies that are frequently employed in investments related to renewable energy (Kumar et al. 2021):

5.1.6.1 Machine Learning Algorithms

- **TensorFlow and PyTorch**: These are machine learning packages that are open source for creating the forecasting models used by the market trends, the performance of the assets as well as energy generated.
- **Scikit-learn**: A Python package providing fast algorithms and with a smooth interaction with the data mining and analysis process. This is often used while designing machine learning models intended for enhancing the investment portfolios as well as intent to identify the future market trends.

5.1.6.2 Natural Language Processing (NLP) Tools

- **Hugging Face Transformers**: With this tool, one is in a position to carry out natural language processing tasks such as the sentiment which can be used to determine the sentiments on the market, social sentiments, or even sentiments on renewable energy projects (Kumar et al. 2024a)
- **Google Cloud Natural Language**: With this API, it is feasible for investors to bring vast quantities of the heterogeneous information about the renewable energy sector, articles, blog posts, research notes, and SMM feeds.

5.1.6.3 Predictive Analytics Platforms

- **IBM Watson Studio**: A sample definition would be—it is an online marketplace for data scientists to train their ML models for better performance. Based on renewable energy resources, the company is capable of predicting future energy prices as well as production output and market currents.
- **SAS Analytics**: It covers a broad area in terms of analytics platform where it offers the parts of predictive analytics and also the part of machine learning. Thus, the aid of an intermediary enables investors to assess the potential of earning revenues from renewable energy assets as well as the situation on the market.

5.1.6.4 AI-Powered Risk Management Tools

- **Riskalyze**: An online tool, which based on algorithms of AI, enables to measure the extent of risk of an investment portfolio. It assists the investor in understanding that there is always some level of risk involved when investing in renewable energy projects and may assist the investor in altering their strategies in such a way that they will not be able to incur losses.

- **Ayasdi**: An analytical platform that uses state of the art algorithms in order to form an analysis of the data. This type is used to evaluate and quantify all forms of risks associated to renewable energy undertakings such as shifts in the policy regime and volatility in energy prices.

5.1.6.5 Energy Production Forecasting Tools

- **NREL's System Advisor Model (SAM)**: A financial and performance model created, whereby, the rationale: It was to provide efficiency in decision-making concerning the renewable energy projects. By using information about the climate in the past, it is able to employ the use of AI in estimating generation potential of solar as well as wind plants.
- **WindSim**: It is one of the software applications that employ the use of AI in estimating energy production and wind resource. It assists in the forecasting of the amount of energy that a wind project will produce; this is an important factor that is used in the financial analysis and other strategic hardcore activities in any investment business.

5.1.6.6 Blockchain and Smart Contracts Platforms

- **Ethereum**: To improve these investments, a blockchain smart contract integrated platform was proposed. Similar to how it can facilitate trading, smart contracts for instance can enhance the legal aspects of investment agreements in the sense that they can be programmed to run as soon as a specific condition is met.
- **Energy Web Chain**: It is another unique and specially developed blockchain only for the energy market. It encourages the decentralized application development or dApps which can quantify energy generation, manage the renewable energy related transactions, and provide transparency over the investing process.

5.1.6.7 Data Analytics and Visualization Tools

- **Tableau**: An approach based on the Russian language has been developed. It is intended for investors and allows them to assess and visualize diverse and complex information regarding the details and characteristics of investments in renewable energy, for example, market landscapes, projected energy outputs, ESG indicators.
- **Power BI**: It serves as a business intelligence application for interactive data visualization and provides simple enough business intelligence reporting tools for users to build reports and dashboards. It is used for monitoring and performing the performance analysis of active investments in renewable energy (Kumar et al. 2023a, b)

5.1.6.8 Energy Storage Optimization Tools

- **Stem, Inc.**: A company that develops AI technologies in the field of storage and energy management. The stem's Athena platform operates in management mode to improve investments in renewable energy by predicting energy usage and controlling the number of charge and discharge cycles of the energy storage systems.
- **AutoGrid**: For this deployment of managed distributed energy resources, the platform is equipped with appropriate energy storage solutions. It helps the investors in enhancing the effectiveness of their energy assets by directing storage or demand response in near real-time.

These AI applications are important for the development of investors' skills in the sector of renewable energy. Utilizing these tools, investors can improve their investment decisions, maximize the returns on their investments, and contribute to the global energy transition (Suresh et al. 2024).

5.2 Literature Review

Investments, especially in renewable energy, are becoming more necessary in the advancement of a low-carbon economy. Investments strategies are being enhanced through the aid of AI which makes it easier to make decisions, assess risks, and monitor impacts (Lim 2024). Keeping in mind a focus on renewable energy, this literature review attempts to explore how AI is applied in implementing sustainable investing policies.

5.2.1 Sustainable Investment in Renewable Energy

- **Sustainable Finance Principles**: As these internal and external factors have come to light, several authors have pointed out the increasing importance of sustainable finance approaches to fund allocation for responsible activities. Investment in the renewable energy sector, which is after all a social venture, is no longer described elsewise by Clark et al. (2015).
- **Impact of Renewable Energy on Sustainable Development**: In works Menyah and Wolde-Rufael (2010) and Apergis and Payne (2010), the authors pay particular attention toward the reason why countries have to consider investing in the renewable energy resources where they stress the fact that investment in clean energy brings about economic sustainability while reducing carbon emissions (Kumar et al. 2024b)

5.2.2 *Role of AI in Sustainable Investment*

- **AI in Financial Decision-Making**: AI transforming the fabric of making financial decisions is a common theme. Dhar (2018) and Choudhry et al. (2018) research how AI-driven models manage risks, optimize portfolios, and forecast market trends in sustainable investments like renewable energy.
- **Machine Learning and Predictive Analytics**: This is to help investors make informative choices. Bergstra et al. (2011) research describes an application of machine learning approaches including predictive analytics to forecast performance pertaining to renewable energy assets.
- **AI in ESG Evaluation**: AI is also applied to improve ESG factor evaluation. A study by Henisz et al. (2019) argues that AI technologies could evaluate firm ESG performance with greater precision compared to traditional techniques, which will improve the alignment of investments with sustainability goals (Kumar et al. 2024c).

5.2.3 *AI in Renewable Energy Market Analysis*

- **Market Forecasting**: Such is the case of Tseng et al. (2020) and Li et al. (2020), in their research studies on the possibilities of AI to forecast developments in the renewable energy industry through the analysis of big datasets that include historical market data, government regulations, and technical improvements.
- **Scenario Planning and Risk Management**: Literature by Agrawal et al. (2018) specifically considers the use of AI in scenario planning for renewable energy investments. Anticipating and reacting to different market conditions and regulatory changes as a result of investment in renewable energy allows an investor to make better decisions.

5.2.4 *Optimization of Renewable Energy Investment Portfolios*

- **Portfolio Management**: The first example, optimization of investment portfolios using AI, is considered by Markowitz (1952) and more recently by DeMiguel et al. (2009). These works demonstrate how AI can adopt portfolio optimum throughout the life cycle based on new information from the market to guarantee alignment between financial objectives and sustainability goals
- **Risk Mitigation**: A second example relates to Bianchini et al. (2019) where it is shown that AI can be used to identify specific risks for renewable energy investments that are not suitable or easily modeled through traditional financial engineering tools like EV integration. More specifically, technological risk

reflects lack of learning-by-doing effect in new technologies, or component failures; regulatory risk refers to changes in tax benefits or grid connection benefits overtime that may affect renewable profitable projects; sectorial risk arises from competition within the same green technology industry segment causing reduced profitability overtime for all participants; interest rate risk is related with cost-of-capital increase due global crises leading investors to reduce their long-term and higher-risk portfolios (Kumar et al. 2024b).

5.2.5 *Challenges and Ethical Considerations*

- **Data Privacy and Security**: Concerns regarding data privacy and security are reviewed in literature by Barocas et al. (2019), who argue the importance of strong data governance structures for the use of AI in financial markets.
- **Algorithmic Bias and Fairness**: Mehrabi et al. (2021) review the literature around bias within AI algorithms, which can lead to unfair investment decisions, and highlight that AI systems should be interpretable and fair.
- **Regulatory Compliance**: Zetzsche et al. (2019) in their works explain the existing legal framework of AI in the finance sector and the challenges that emerge in the sphere of how using AI to develop investment strategies is able to offer adequate solutions to the various financial legislation systems.

The analysis of literature proves that AI affects sustainability by improving the decisions, improving the ESG evaluation, and improving the portfolio. Nevertheless, should the question of data privacy arise, the question of explaining the effects of the biases of an algorithm, or the question of adherence to the regulation, they remain.

Therefore, it is possible to describe several more specific directions for the further research which consist in studying the application of the AI along with the other potentially effective technologies, for example, blockchain or IoT in the context of investments in renewable energy. This has given rise to questions regarding the social and environmental impact of AI-driven investment decisions that also calls for more research in those fields.

5.3 The History and Role of AI in Sustainable Investment Policies in Renewable Energy

5.3.1 *Early Development: Pre-2000s*

The concept of sustainable investment began gaining traction in the 1980s and 1990s, focusing primarily on divesting from fossil fuels and investing in green energy initiatives. But these investment decisions were normally taken with almost complete disregard to most forms of quantitative reasoning and methods of valuation except those involving basic deterministic calculations.

Some of the earliest models of investments were built on historical sales data, and this used arithmetic analysis and then the screen of the impact of the environment on investment opportunities. Since there was no high level of development of the ideal instruments, there was limited capability to examine and evaluate the various risks and the returns on investments in renewable energy sources.

This was especially true in the areas of finance and sustainability where automation using AI was almost non-existent and this was because at this time AI was still in its infancy (Suresh et al. 2024).

5.3.2 The Emergence of AI in Sustainable Investment: 2000s

The usage of AI applied to the adoption of sustainable investment policies began at the early 2000s including renewable energy.

With the rise of more advanced data analytics and AI techniques, investors began to incorporate quantitative data into their decision-making processes. New environmental, social, and governance (ESG) data, which was enabled by AI, contributed to the recognition of possible investment in the renewable energy sector.

The efficiency of AI algorithms was used in the assessment of the financing risks that are associated with renewable energy undertakings. This entailed identifying performance trends of energy generation, the market position, and the legal environment to be able to identify performance and possible fluctuations in the future.

Socially responsible investment (SRI) is established to have evolved from the 2000s when ESG metrics was also established to have begun its evolution with backing from AI analytics.

5.3.3 AI in Sustainable Investment: 2010s

In the second decade of the twenty-first century, AI can be identified as one of the factors that have an impact on the creation of sustainable investment politics for renewable energy.

AI started to play a crucial role in integrating ESG factors into investment strategies. Applying big data approaches, big data was gathered particularly on the environmental sustainability reports, social responsibility stock, governance factors, and reviewing big data on that identified high potential renewable energy investments.

AI began to be utilized in the green bonds that was being used in the issuance and the management of instruments in the green bond market that is utilized in the funding worthy projects within the environmental sectors such as the renewable power projects. The green bonds used AI models to determine sustainability impacts of the projects that are funded and also verify compliance with green metrics.

The latter part of the 2010s also saw the debut of other AI-based platforms that offered indexes to enhance investors insight into the sustainability of industries such as renewable energy to enable them to invest far more strategically (Ahmad et al. 2021).

5.3.4 AI in Sustainable Investment: 2020s

AI has become even more engaged in sustainable investment policies at the onset of 2020s with more focus and specialization in renewable energy investment policies because of a rising global interest coupled with the ever-evolving AI technologies.

Employing AI is not an option for companies in the modern world since it is instrumental in the elaboration of ESG data to enable investors make correct estimations on the sustainability performance of renewable energy companies. For instance, natural language processing (NLP) where the algorithms found useful in scrutinizing the company reports, news articles, and social media to approximate the ESG risks and opportunities.

AI is promoting evolution of impact investment that seeks to bring value addition on social and environmental fronts in addition to financial returns. To the investors, assistance in project selection based on the testable AI algorithms is injection sufficient for making right choices within the field of renewable power both from the investment engagement stand as well as environmental standards including low-carbon projects.

AI is used as a matter of routine in determining the climate risks that shall influence the investment in renewable power sources. For instance, the models can identify how some situations like a calamity, increase in sea level, or change of climate will affect the construction of recrual renewable energy or the feasibility of a specific project.

There is the emergence of AI-based applications that offer UI solutions for sustainable investments that entail a method for screening, tracking, and reporting on investments in renewable energy. It enables platforms that can provide investors with a vision of sustainability performance of their investments, and also help them adjust their holdings to the environmental characteristics.

Today we can witness how AI can help individuals to invest efficiently in the transition to sustainable investment to invest in renewable energy projects in the 2020s.

5.4 Key Challenges of AI in Sustainable Investment Policies in Renewable Energy

While AI is clearly going to play a bigger role than ever in sustainable investment policies under the rubric of renewable energy growth, challenges remain. These challenges affect technology, ethics, and regulation and have major impact on functionality as well as responsible use of AI in that area (Chen et al. 2021).

5.4.1 Data Quality and Availability

Decisions made by AI models are heavily dependent on the availability of large, structured data. Sometimes the information is patchy for investing in renewable energy applications that are long term and sustainable. For instance, the environmental/social/governance (ESG) data necessary in evaluating the sustainability of an investment is broadly absent standardization here with individual companies selecting what to disclose. This complexity results in increased difficulty for AI models to correctly assess and compare the renewable energy projects that are available.

The availability and accessibility of real-time data needed by AI for decision-making within dynamic ecosystems, such as energy markets, may be restricted, e.g., real-time energy production, market prices, or weather data preventing timely and accurate recommendations.

5.4.2 Complexity of ESG Criteria and Standards

ESG criteria are at the core of sustainable investment; however, there is a lack of standardized metrics and reporting frameworks within the renewable energy sector, where different organizations and jurisdictions may integrate different criteria for sustainability assessment, leading to discrepancies in data applied to AI models.

The standards and regulation for ESG are constantly changing, developing, and being improved upon. This makes it difficult for AI systems to keep up. Updating an AI model with the most recent ESG criteria is labor intensive both in terms of costs and use of resources.

5.4.3 Regulatory and Governance Challenges

As important at this stage is the fact that there are still many jurisdictions in the process of adopting changes to the rules and standards of application of AI in way for sustainable investment. The kinds of predictability that can be this problematic for both investors and the financial institutions are volatile regulatory environments.

There is a broad range of approaches that one country or another, regulators and Standards Setters, apply while building up regulation and Standards of AI, ESG criteria, and sustainability investment. For investors having licenses in multiple jurisdictions, it becomes a problem to integrate the AI strategies as well as always rousing compliance questions.

5.4.4 Long-Term Impact and Uncertainty

Anticipating the returns of renewable energy investments is bound to be full of risk and uncertainty especially with regard to future growth of the technologies and changes in the political and legal structures. Proposed AI models can fail to consider long-term conditions that may affect the risk and return of investment hence making wrong decisions (Narayanan et al. 2021).

Climate change brings uncertainty that is likely to impact the performance of renewable energy projects. Although some of these risks can be modeled with the help of AI, the nature of climate change is such that it is hard to predict future results; as a result, it is possible that investments will be improperly aligned.

As earlier, this sector is characterized by the adoption of new technologies whose growth rate is very fast hence catching up with the technologies available in the market is a challenge. It implies that AI models have to evolve in response to such changes, so as to keep the choice of investment solutions valuable and efficient.

5.5 Sustainable Investment Policies in Renewable Energy Through AI

Use of AI in investment to renewable energy is a rising subject, as many governments, financial institutions, and firms have been trying to enhance the efficiency of energy and embraced shift toward low-carbon economy in the world presently (Şerban and Lytras 2020). Here is a list of sustainable investment policies that are being implemented or considered to promote renewable energy through AI:

5.5.1 Green Finance and ESG Integration

The AI in assessing and monitoring the environmental, social, and governance (ESG) performance in checking on the investments on renewable energy to ensure they have met set standards of sustainability.

Automatization of project certifications of renewable energy for admission of green financial instruments to assist investors in financing concession projects with impact on the environment.

5.5.2 AI-Powered Risk Assessment and Management

The feasibility analysis for the possible risks associated with renewable energy investments such as market change, change in legal bound, or climate change, among others, when employed with machine learning and deep learning algorithms.

5.5.3 Government Incentives and Tax Credits

Using of AI for the control of the limited resources of the government subsidies for renewable energy projects. To offer such encouragements like tax credits to those organizations that integrate the AI applications in renewable energy to make them effective and sustainable.

5.5.4 Carbon Pricing Mechanisms

Utilizing AI in the sphere of carbon pricing—the actual price applied to carbon dioxide and other pollutants, the carbon emissions trading and other emissions trading schemes, for direct investments accord for equipment and infrastructure, and other fundamental low-carbon technologies, and renewable energy systems. Automated carbon footprint monitoring means the use of AI in tracking and realizing the carbon goals of energy projects that may in turn shift investment strategies and guidelines.

5.5.5 Smart Grid Investments

Backing the AI-based smart grids investment which enables high degrees of using renewable energy, efficiency, and reducing the operational cost. They utilize AI in controlling the amounts of energy demanded in relation to supply conditions and the renewable energy resources are well controlled.

5.5.6 Public–Private Partnerships (PPP)

PPP helps the individuals within the governmental agencies, private enterprises, and the AI developers collaborate and make investment in renewable energy solutions and energy infrastructure.

5.6 Sustainable Investment Policies in Renewable Energy Versus Sustainable Investment Policies in Renewable Energy Through AI

The first kind of programs is sustainable investment policies in renewable energy and the second is sustainable investment policies in renewable energy through AI are programs that work to help transform the economy to rely more on sources of renewable energy and in the process develop the economy to be more sustainable. But they are in terms of the area of interest and the approach that is taken to deal with it. Following are the comparisons:

5.6.1 Scope and Focus

- Renewable Energy Policies: It encourages investments in renewable energy technologies, such as solar, wind, hydro, and biomass, with a broad focus and the use of clean energy sources, improve energy security, and lower carbon emissions.
- Renewable Energy Through AI Policies: It pays close attention to how you may use AI technology to maximize, improve, and stimulate renewable energy investments and improve decision-making, risk management, predictive analytics, and efficiency in renewable energy projects by utilizing AI.

5.6.2 Key Objectives

- Renewable Energy Policies: It boosts the proportion of renewable energy in the energy mix overall and cut back on your reliance on fossil fuels. It promotes creativity and brings down the cost of renewable energy sources.
- Renewable Energy through AI Policies: It increases the precision of project outcomes and forecasts for renewable energy by utilizing AI and boosts the dependability and efficiency of renewable energy systems. It mitigates the hazards linked to investments in renewable energy by enhancing data analysis and predictive modeling.

5.6.3 Mechanisms and Tools

- Renewable Energy Policies: It incorporates conventional policy instruments such as grants, feed-in tariffs, tax incentives, subsidies, and renewable portfolio standards (RPS) and encourages the construction of infrastructure for renewable energy sources through public–private partnerships and government subsidies.
- Renewable Energy Through AI Policies: It includes AI-powered solutions like automated project optimization, smart grid management, dynamic pricing models, and predictive analytics and develop AI applications, such as AI-powered energy storage systems or AI-based grid management, that can be incorporated into renewable energy systems.

5.6.4 Implementation Strategies

- Renewable Energy Policies: It is implemented by governmental mandates, financial incentives, and regulatory frameworks to promote the use of renewable energy and frequently involve widespread backing for the advancement of renewable technology research and development.
- Renewable Energy through AI Policies: It is implemented by promoting funding for AI invention and research that is especially suited to the renewable energy sector and may entail the development of guidelines and frameworks for AI applications in the energy sector, along with financial rewards for initiatives that leverage AI.

5.7 Conclusion

The incorporation of AI in renewable energy sector has provided solutions to issues of production, management, and consumption of energy. AIIs are crossing through virtually all aspects of the energy transition, ranging from modern wind and solar generation to intelligent distribution networks and energy storage. As a result, renewable energy industry will continue its growth through the assistance of AI technologies and add more values in decreasing the effects of climate change.

Funding NA.

Conflict of Interest There is no conflict of interest for this research.

References

Ahmad T, Zhang D, Huang C, Zhang H, Dai N, Song Y, Chen H (2021) Artificial intelligence in sustainable energy industry: status Quo, challenges and opportunities. J Clean Prod 289:125834

Agrawal A, Gans JS, Goldfarb A (2018) Human judgment and ai pricing. In: AEA Papers and Proceedings vol 108. 2014 Broadway, Suite 305, Nashville, TN 37203: American Economic Association, pp 58–63

Apergis N, Payne JE (2010) Renewable energy consumption and economic growth: evidence from a panel of OECD countries. Energy Policy 38(1):656–660

Barocas JA, Wang J, Marshall BD, LaRochelle MR, Bettano A, Bernson D, Beckwith CG, Linas BP, Walley AY (2019) Sociodemographic factors and social determinants associated with toxicology confirmed polysubstance opioid-related deaths. Drug Alcohol Depend 200:59–63

Bergstra J, Bardenet R, Bengio Y, Kégl B (2011) Algorithms for hyper-parameter optimization. Adv Neural Inf Process Syst 24

Bianchini S, Llerena P, Patsali S (2019) Demand-pull innovation in science: empirical evidence from a research university's suppliers. Res Policy 48:100005

Chen C, Hu Y, Karuppiah M, Kumar PM (2021) Artificial intelligence on economic evaluation of energy efficiency and renewable energy technologies. Sustain Energy Technol Assess 47:101358

Choudhry M (2018) An introduction to banking: principles, strategy and risk management. John Wiley & Sons

Clark GL, Feiner A, Viehs M (2015) From the stockholder to the stakeholder: how sustainability can drive financial outperformance. Available at SSRN 2508281

Danish MSS, Senjyu T (2023) AI-enabled energy policy for a sustainable future. Sustainability 15(9):7643

DeMiguel V, Garlappi L, Uppal R (2009) Optimal versus naive diversification: how inefficient is the 1/N portfolio strategy?. Rev Financ Stud 22(5):1915–1953

Dhar S (2018) Gender and sustainable development goals (SDGs). Indian J Gend Stud 25(1):47–78

Ferdaus MM, Dam T, Anavatti S, Das S (2024) Digital technologies for a net-zero energy future: a comprehensive review. Renew Sustain Energy Rev 202:114681

Hannan MA, Al-Shetwi AQ, Ker PJ, Begum RA, Mansor M, Rahman SA, Muttaqi KM et al (2021) Impact of renewable energy utilization and artificial intelligence in achieving sustainable development goals. Energy Rep 7:5359–5373

Henisz W, Koller T, Nuttall R (2019) Five ways that ESG creates value. McKinsey Q 4:1–12

Kou G, Pamucar D, Dinçer H, Yüksel S (2023) From risks to rewards: a comprehensive guide to sustainable investment decisions in renewable energy using a hybrid facial expression-based fuzzy decision-making approach. Appl Soft Comput 142:110365

Kumar P, Singh SK (2017) A framework for assessing the evolvability characteristics along with sub-characteristics in AOSQ model using fuzzy logic tool. In: 2017 International conference on computing, communication and automation (ICCCA). IEEE, pp 339–344

Kumar P, Singh, S. K. (2020). An emerging approach to intelligent techniques—soft computing and its application. In: Internet of Things and Big Data applications: recent advances and challenges, pp 171–182

Kumar P, Sharma R, Choudhary SD, Singh SK, Kapse VM (2021) Predictive analysis of novel coronavirus using machine learning model-a graph mining approach. J Math Comput Sci 11(3):3647–3662

Kumar R, Kandpal B, Ahmad V (2023a) Industrial IoT (IIOT): security threats and countermeasures. In: 2023 International conference on innovative data communication technologies and application (ICIDCA). IEEE, pp 829–833

Kumar R, Malholtra RK, Grover CN (2023b) Review on artificial intelligence role in implementation of Goods and Services Tax (GST) and future scope. In: 2023 International conference on artificial intelligence and smart communication (AISC). IEEE, pp 348–351

Kumar M, Rai PK, Kumar P (2024a) A novel approach for detecting deepfake face using machine learning algorithms. In: 2024 2nd International conference on disruptive technologies (ICDT). IEEE, pp 1588–1592

Kumar R, Goel R, Singh T, Mohanty SM, Gupta D, Alkhayyat A, Khanna R (2024b) Sustainable finance factors in Indian economy: analysis on policy of climate change and energy sector. Fluct Noise Lett 23(2)

Kumar V, Kumar R, Goel R, Singh T, Kaur MJ (2024c) The evolution of digital marketing technologies: a fuzzy TOPSIS approach to selected technology in E-commerce platform. Int J Syst Assurance Eng Manage:1–14

Li C, Yang Y, Ren L.(2020) Genetic evolution analysis of 2019 novel coronavirus and coronavirus from other species. Infect Genet Evol 82:104285

Lim T (2024) Environmental, social, and governance (ESG) and artificial intelligence in finance: state-of-the-art and research takeaways. Artif Intell Rev 57(4):76

Markowitz H (1952) Modern portfolio theory. J Financ 7(11):77–91

Mehrabi Z, Ramankutty N (2019) Synchronized failure of global crop production. Nat Ecol Evol 3(5):780–786

Menyah K, Wolde-Rufael Y (2010) CO2 emissions, nuclear energy, renewable energy and economic growth in the US. Energy Policy 38(6):2911–2915

Narayanan G, Ali MS, Alam MI, Rajchakit G, Boonsatit N, Kumar P, Hammachukiattikul P (2021) Adaptive fuzzy feedback controller design for finite-time Mittag-Leffler synchronization of fractional-order quaternion-valued reaction-diffusion fuzzy molecular modeling of delayed neural networks. IEEE Access 9:130862–130883

Selim O (2020) ESG and AI: the beauty and the beast of sustainable investing. In: Sustainable investing. Routledge, pp 227–243

Şerban AC, Lytras MD (2020) Artificial intelligence for smart renewable energy sector in Europe—smart energy infrastructures for next generation smart cities. IEEE Access 8:77364–77377

Suresh C, Nyemeesha V, Prasath R, Lokeshwaran K, Raju KR, Boopathi S (2024) AI-driven energy forecasting, optimization, and demand side management for consumer engagement. In: Intelligent solutions for sustainable power grids. IGI Global, pp 112–131

Tseng ML, Chang CH, Lin CWR, Wu KJ, Chen Q, Xia L, Xue B (2020) Future trends and guidance for the triple bottom line and sustainability: a data driven bibliometric analysis. Environ Sci Pollut Res 27(27):33543–33567

Zetzsche DA, Buckley RP, Arner DW, Fohr L (2019) The ICO gold rush: It's a scam, it's a bubble, it's a super challenge for regulators. Harv. Int'l LJ:60:267

Chapter 6
Financial Inclusion Through Fintech: Bridging the Gap

V. Harish, Ravindra Sharma, Arun Karthik, Geeta Rana, R. D. Balaji, and Rakesh Kumar

Abstract Financial inclusion, or equal access to financial services, is crucial to global economic progress. This extensive study examines how financial technology (fintech) advances this noble cause. The introduction sets the stage for our study of fintech's transformational potential in financial inclusion. The literature study explains fintech's history and complex relationship with financial inclusion. Our analysis is set against the backdrop of fintech technologies' transformation of the financial services sector. The next part discusses fintech developments like mobile banking and blockchain and their substantial benefits in decreasing financial inclusion barriers. We examine how new technologies have increased financial access, especially for marginalized people. The authors also aim to identify and analyse financial exclusion barriers such as lack of financial services, literacy, and infrastructure impacting equal access for all. The study also explains how financial inclusion boosts economic growth, poverty alleviation, and social development beyond morality. Digital payment systems, peer-to-peer lending, and new credit scoring algorithms are examples. We recognize fintech's implementation issues and concerns, but its potential is promising. Regulatory frameworks, cybersecurity hazards, and data

V. Harish · A. Karthik
PSG Institute of Management, PSG College of Technology, Coimbatore, India
e-mail: harish@psgim.ac.in

A. Karthik
e-mail: arunkarthik@psgim.ac.in

R. Sharma (✉) · G. Rana
School of Management Studies, Swami Rama Himalayan University, Dehradun, India
e-mail: ravindrasharma@srhu.edu.in

G. Rana
e-mail: geetarana@srhu.edu.in

R. D. Balaji
Department of IT, University of Technology and Applied Sciences (Salalah-CAS), Saadha, Sultanate of Oman
e-mail: balaji.sal@cas.edu.om

R. Kumar
Department of Management Studies, Graphic Era (Deemed to be) University, Dehradun, India

R. Kumar (ed.), *Sustainable Finance*, Sustainability Solutions,
https://doi.org/10.1007/978-3-031-97962-0_6

privacy are important here. Our study examines how governments promote financial inclusion. The paper also studies impact evaluation and measurement methods to evaluate financial inclusion activities. The next section lists metrics for measuring financial inclusion programmes. While the impact assessment is difficult, measures provide vital insights. Complexities including context, data restrictions, and long-term consequences must be considered. Finally, our research shows how fintech supports the sustainable development goals. We examine how fintech-driven financial inclusion affects sustainable development. This report provides a detailed overview of fintech and financial inclusion's dynamic relationship. It highlights fintech's disruptive promise while navigating its intricacies and pitfalls. We have vital insights to improve financial inclusion in our ever-changing financial landscape as researchers, educators, policymakers, and practitioners.

Keywords Fintech · Financial inclusion · Sustainable development goals · Barriers · Technology

6.1 Introduction

The idea of "financial inclusion" has gained momentous importance in today's interconnected, technologically advanced and developed world (Govindapuram et al. 2023). Traditional financial systems for long have essentially served a large and significant portion of the population, but they have left out a significant segment of people worldwide from being in a position to access the various benefits provided by formal banking and financial service systems (Maity and Sahu 2022). This act of excluding people popularly known as financial exclusion not only attributes to the loss of critical financial resources, but also more importantly acts as a barrier to the overall progress and development of the country. Governments and institutions have constantly aimed to address the problem of financial inclusivity (Chen et al. 2021). One of the promising solutions to address this issue at hand is the adoption of various technologies that shifts gears from the traditional financial systems. The onset and rapid growth of various technologies in the financial industry is popularly known as fintech and offers hope for institutions and governments to solve the issue of financial inequality. Figure 6.1 explains the grouping of the papers considered for this study.

The rapid evolution of fintech over the years, based on the mix of "finance and technology", has made a significant shift towards the use of innovative technologies and methods targeting to bridge the gap between those who have access to financial services and those who do not (Bollaert et al. 2021; Prakash et al. 2021). The purpose of this study is to delve into the concepts of fintech and financial inclusion through an extensive literature review, followed by an exploration of the various definitions of fintech and to examine fintech technologies, highlighting their benefits in relation to financial inclusion, while also addressing the challenges that come with achieving financial inclusion and the pressing need to tackle issues of exclusion. The next part

Fig. 6.1 Literature review—topic-wise breakup of articles

addresses the various fintech solutions for financial inclusion and challenges and concerns of fintech solutions for financial inclusion. The next part of the chapter discusses the part of government in endorsing financial inclusion and impact assessment and measurement. The last part of the study attempts to address fintech's role in sustainable development goals.

Table 6.1 indicates a sample of the literature review done for the study.

Different interpretations of financial inclusion have been formulated based on a thorough evaluation of available literature. The following discussion showcases a variety of these ideas.

- Improved Financial Accessibility: Fintech ensures accessibility of various financial products and services, including banking services, savings options, credit opportunities, insurance offerings, and payment systems, to all segments of society. This includes a focus on historically marginalized or underserved groups (Ozili 2020).
- Promoting Economic and Social Inclusivity: Financial inclusion is not just about offering financial services to all but should also ensure that the economic and social well-being of people and communities (Ediagbonya and Tioluwani 2023).

The above-mentioned definitions highlight the varied and diverse nature of the concepts of financial inclusion and fintech, which focuses delivery of financial services and economic growth.

I. **Financial technology technologies and their benefits for financial Inclusion**

Table 6.1 Some of the prominent studies (indicative list)

Author	Citation	Abstract summary	Outcome measured	Region	Detailed study type
Rhanoui (2023)	0	"Fintech is on the right track to promote financial inclusion through bank cards"	• ATM card as a fintech tool •Bank cards in circulation for six years •Potential contributions of fintech industry to greater financial inclusion	–	Forecast scenario analysis
(Datta, 2023)	1	Fintech will assist the governments in improving the economy post-pandemic	Provide a understanding on how financial inclusion can be achieved through India's post-covid	India	Descriptive
Joia and Cordeiro (2021)	20	Fintechs can • Serve niches people • Serve people engaged in traditional financial market	• How fintechs' can assist in financial inclusion •How fintechs can cater to the needs of people in remote locations thereby including them in financial development	Brazil	Delphi technique
Demir et al. (2022)	259	• Financial inclusion vital • Fintech key component in achieving inclusion • Fintech technology adoption can reduce income inequality	• Financial inclusion • Reduction of income inequality	–	A panel study
Senyo and Osabutey (2020)	192	Intention to use digital money services is highly influenced by • Performance of the technology and • Effort expectancy required	• Intention to use digital money services • Adoption of mobile money	Ghana	Survey based
Gabor and Brooks (2020)	668	Digital revolution and adoption leads to financial inclusion	Growing reputation of digital-based financial inclusion leads to overall development	–	–

(continued)

Table 6.1 (continued)

Author	Citation	Abstract summary	Outcome measured	Region	Detailed study type
Mehrotra (2019)	26	Fintech redefines • Service quality • Better delivery • Reduces inequality	• Financial inclusion • Reduced inequality • Service quality	–	Literature review
Makina (2019)	72	• Mobile money in Africa • How it assists financial inclusion	•Impact of selected fintech products • Impact of mobile money •Challenges for adoption of fintech	Africa	–

See Table 6.2.

6.2 Barriers to Financial Inclusion

The barriers are indicated as shown in Fig. 6.2.

1. **Economic Disparities**: Wide income gaps can hinder access to financial services, as those with lower incomes often struggle to meet minimum account requirements and lack of wealth or assets can act as a barrier which can prevent people getting loans due to lack of collaterals and reduce eligibility for financial products (Liu et al. 2022)
2. **Limited Financial Literacy**: (Kakinuma 2023) Insufficient knowledge and lack of knowledge about financial concepts, financial technology, and products can act as a barrier for individuals to deal and engage with formal financial institutions and generally individuals are scared with the overly complex jargons and terminologies, which often create fear, apprehension, confusion, and usually discourage people from using financial services/products (Kakinuma 2023).
3. **Geographic Accessibility**: In developing countries like India rural areas often lack physical bank branches and ATMs, forcing residents to travel long distances for basic services which acts as a big barrier for financial inclusion and inaccessibility due to geographical barriers, such as mountains or islands, can isolate communities and societies from financial infrastructure and often are financially excluded unlike the mainstream population (Ezzahid and Elouaourti 2021).
4. **Documentation Requirements**: Stringent identity verification processes in the name of validation of data and adherence to policies and procedures of financial institutions and government polies can be a hurdle, especially for marginalized populations without proper and adequate identification and this can lead to financial exclusion of the under privileged (Huibers 2021).

Table 6.2 Fintech technology and financial inclusion benefits

Fintech technology	Benefits for financial inclusion
Mobile banking apps	1. Banking services via smartphones
	2. Reduce visit to banks
	3. Facilitates easy fund transfers and payments
	4. Increases the accessibility of services
	5. Helps in reaching remote and underserved areas
	(Putritama 2019)
Digital wallets	1. Offers a convenient and secure way to store money
	2. Supports peer-to-peer transactions for the unbanked
	3. Reduces the reliance on cash transactions
	4. Enhances access to online marketplaces for purchases
	(Anshari et al. 2021)
Blockchain technology	1. Provides transparent and immutable transaction records
	2. Enables cross-border remittances at lower costs
	3. Reduces fraud and identity theft risks
	4. Expands financial services to the unbanked
	(Kumari and Devi 2022)
Peer-to-peer lending	1. Allows individuals to borrow and lend without banks
	2. Offers lower interest rates to borrowers
	3. Connects lenders with borrowers directly
	4. Increases access to credit for underserved populations
	(Rusadi and Benuf 2020)
Biometric authentication	1. Enhances security and reduces fraud in transactions
	2. Provides access to financial services via
	(Goyal and Chakrabarti 2022)

5. **High Transaction Costs**: Excessive fees for basic transactions, such as withdrawals and transfers, can erode the value of small savings and these type of costs can act as a deterrent for the people to engage in the financial inclusion process and other additional costs such as the cost of travelling in order to access the mentioned financial services or fintech products can significantly outweigh the benefits (Alkhazaleh and Haddad 2021).
6. **Digital Divide**: Inadequate or lack of Internet / digital infrastructure and poor network coverage can prevent people from using financial digital products and generally causes a digital divide and this can be further aggravated by lack of digital Literacy skills can make it cumbersome for people to use recent digital financial tools for better reach and upliftment (Odei-Appiah et al. 2022).
7. **Regulatory Barriers**: Strict regulations from regulatory authorities can be discouraging factor for financial institutions from offering services to specific segment and policies governing to anti-money laundering (AML) requirements

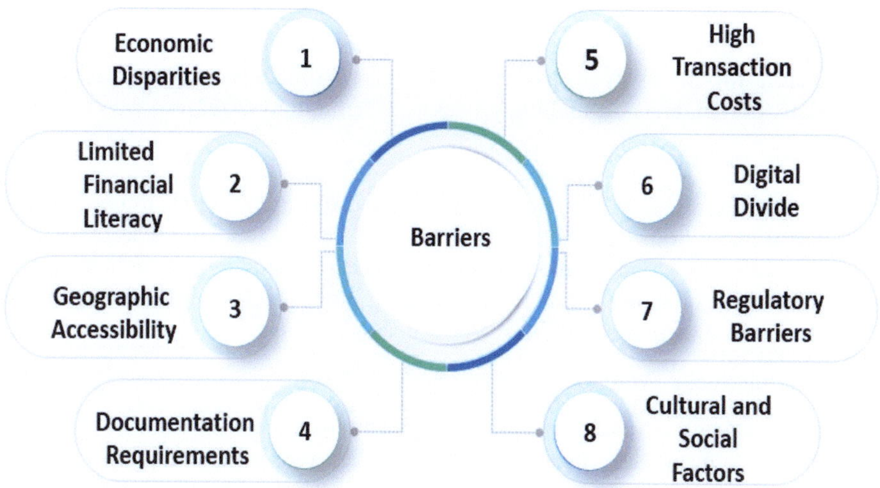

Fig. 6.2 Barriers to financial inclusion

that are cumbersome and complicated AML compliance measures can deter banks and financial institutions from serving low-income customers (Imam et al. 2022).

8. **Cultural and Social Factors**: Cultural aspects/beliefs and social norms may discourage certain groups, such as women or specific communities, from engaging with formal financial institutions and other factors such as gender disparities can act as a constraint for women's financial autonomy and restrict access to financial services/financial institutions (Cambaza 2023).

6.3 Significance of Addressing Financial Exclusion

Some of the significant points that point out the importance of addressing tackling the issue of financial exclusion are given below:

Economic Empowerment:

- Financial inclusion enables marginalized and low-income populations to access the basic offerings of savings and credit and this can help people to move out of poverty and live a better life and expanding financial inclusion through adopting fintech products fosters economic growth by broadening the base of users thereby stimulating consumption and in turn, promoting prosperity (Zuliansyah et al. 2022).

Reducing Inequality:

- Financial inclusion serves as a tool for wealth redistribution by providing equal access for all segments of people to access financial resources and promoting inclusion for all as it empowers women economically and increases their participation in decision-making processes (Demir et al. 2022).

  Strengthening Financial Stability:

- A comprehensive and all-inclusive broader and more diverse customer base for financial institutions can enhance financial stability by reducing risks associated with concentrated lending and investments and earlier studies have indicated individuals with inclusion are safe from economic shocks and crises by having access to savings and insurance products (Jameaba 2020).

  Enhancing Social Welfare (Panos and Wilson 2020):

- Access to fintech can enable people by helping them manage unexpected expenses, plan for healthcare and education, and secure their future and acts as social safety nets can assist delivery of government subsidy to the socially required more efficiently (Panos and Wilson 2020).

  Promoting Digitalization:

- The growth of financial inclusion often goes hand in hand with embracing digital financial services and this promotes technological advancements, encourages innovation in the financial sector, and brings remote and underserved populations into the digital economy and expanded access encourages development which can benefit consumers (Rauniyar et al. 2021; Harish and Sharma 2024).

  Global Development Goals (Hoang et al. 2022):

- Addressing financial exclusion aligns with several SDGs, including "no poverty (SDG 1), gender equality (SDG 5), reduced inequalities (SDG 10), and partnerships for the goals (SDG 17)".

These points specify the role of the financial inclusion through adopting fintech solutions.

6.4 Fintech Solutions for Financial Inclusion

Fintech solutions can be an effective tool for bringing financial inclusion for all by adopting various technologies. Some of the solutions are as shown in Fig. 6.3.

1. Digital Payments and Mobile Wallets:

Accessibility: Digital payment solutions such as mobile wallets can ensure that financial services are available to individuals in remote and underserved areas and thereby improving their financial well-being and financial inclusion and they also provide affordability as digital payments are often a cost-effective alternative to traditional

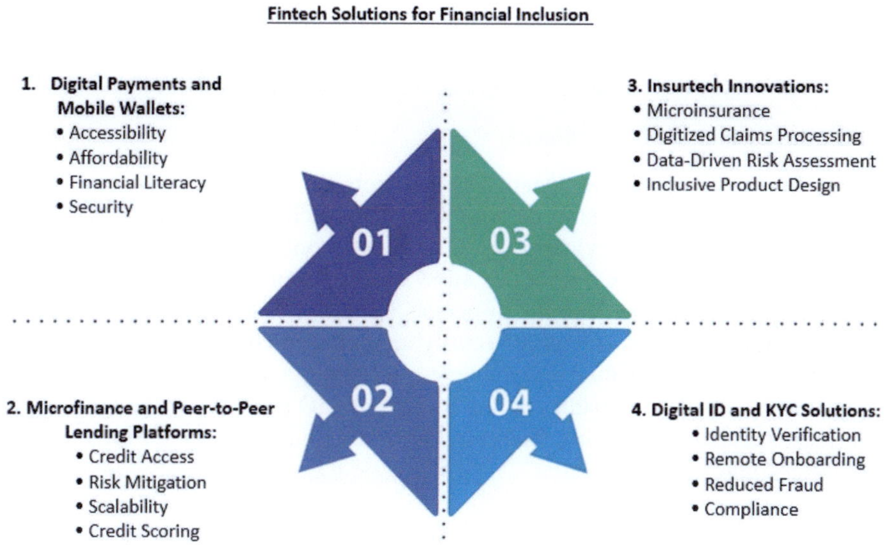

Fig. 6.3 Fintech solutions for financial inclusion

banking solutions and considerably reduces effort as well as it provides convenience to the users by reducing the transaction time and transaction cost considerably. They also improve financial literacy as mobile wallets have extremely user-friendly interfaces and are easy-to-use thereby ensuring financial literacy among users and also ensure security as fintech companies due to regulatory dimensions employ robust encryption and authentication measures to ensure safe digital transactions (Almuhammadi 2020).

2. Microfinance (Sunardi et al. 2022):

Credit Access: Fintech in the form of microfinance platforms offer small loans to underserved segment, thereby allowing them to live better and also can assist in effective risk mitigation where fintech products can act as a risk mitigating tool for users by offering money to the under privileged while also offering scalability where these platforms can efficiently scale their operations and reach previously unreached populations by leveraging technology and data analytics and also can assist in credit scoring as fintech's employ alternative credit scoring models, incorporating new technologies and expanding access to all (Sunardi et al. 2022).

3. InsurTech Innovations:

Microinsurance: Fintech-driven InsurTech solutions offer affordable microinsurance products providing financial protection against risks like illness, accidents, and crop failure and innovations such as digitized claims processing through mobile apps and digital platforms streamlines the insurance process, reducing paperwork and enabling faster payouts and can also assist in data-driven risk assessment where InsurTech leverages data analytics to assess risks more accurately, resulting in fairer premiums and increased affordability for policyholders and can assist in creating inclusive product design where fintech companies work on developing insurance products that meet the requirements and address vulnerabilities (Bittini et al. 2022).

4. Digital ID and KYC Solutions:

Digital ID solutions enable identity verification for individuals without traditional identification documents to establish their identity electronically, crucial for accessing financial services and allow remote onboarding as fintech-driven know your customer (KYC) processes allow users to complete identity verification remotely, eliminating the need for in-person visits to brick-and-mortar banks and result in reduced fraud as enhanced security measures and biometric authentication reduce the risk of identity theft and fraudulent account openings and assist in various compliance as fintech companies ensure their digital ID and KYC solutions comply with local regulations, making them a reliable and compliant option for financial institutions (Ghozi 2022).

6.5 Challenges and Concerns of Fintech Solutions for Financial Inclusion

The challenges and concerns in greater detail with main points and sub-points are as shown in Fig. 6.4.

1. Privacy and Data Security (Mehrban et al. 2020):

   Data Vulnerability:

- Fintech companies collect and process vast amounts of user data, making them potential targets for data breaches and cyberattacks and users may have concerns about how their financial and personal data is stored, shared, and protected and with regulatory compliance regarding data privacy can be complex, with different standards in various jurisdictions (Sampat et al. 2023).

  User Consent:

- Balancing data collection for credit assessment and maintaining user privacy can be challenging. Users must trust that their data is used responsibly and informed consent is essential. Users should be fully aware of how their data will be used and have the ability to control its sharing and building trust through transparent data practices is crucial to overcome privacy concerns (Nelaturu et al. 2022).

  Regulatory Compliance:

- Fintech firms operate in a global landscape, facing diverse data protection regulations that can be challenging to navigate by complying with regulations like GDPR in Europe or CCPA in California requires substantial resources and ongoing efforts (Rabbani 2022).

Fig. 6.4 Challenges of fintech solutions for financial inclusion

2. Regulatory and Compliance Issues (Naz et al. 2022):

Diverse Regulatory Landscapes: Fintech companies in multiple countries follow different rules, posing a complex compliance challenge and working towards anti-money laundering (AML) in different jurisdictions is resource-intensive and changes in regulations can impact business models and require significant adjustments (Mehrotra and Menon 2021).

Consumer Protection:

- Ensuring robust consumer protection standards within fintech platforms is essential and users must have recourse in case of disputes or issues and clear and fair terms and conditions, as well as transparent fee structures, are crucial to protect consumers and adequate redress mechanisms should be in place for users to address grievances (Yuniarti and Rasyid 2020).

Changing Regulations:

- Fintech is an evolving industry, and regulations may struggle to keep up with technological advancements and regulatory changes can create uncertainty for fintech firms, affecting their strategies, product offerings, and market entry decisions and adapting to new regulations and staying compliant requires ongoing vigilance and resources (Omarova 2020).

3. Digital Divide and Access Barriers

Inadequate Internet infrastructure in rural and remote areas limits access to digital financial services, hindering financial inclusion efforts and uneven network coverage and connectivity issues can disrupt the user experience and discourage adoption and reliable and affordable Internet access is fundamental to bridging the digital divide (Cambaza 2023).

Technological Literacy:

- Low levels of digital literacy, particularly among older or less-educated populations, can impede the effective use of fintech services and navigating digital platforms, understanding financial apps, and ensuring security may be challenging for some users and educational initiatives are needed to enhance digital skills among underserved populations (Mhlanga 2022).

Affordability:

- The cost of smartphones, data plans, and digital devices can be prohibitively high for many individuals in low-income or marginalized communities and affordability barriers prevent access to fintech services, exacerbating financial exclusion and innovative approaches, such as affordable smartphones and data subsidies, can help address this challenge (Kukreja 2021).

4. Overcoming Trust Deficits (Mehrotra 2019):

   Lack of Trust: Traditional financial institutions have established trust over time, while fintech startups may face scepticism from users and building trust requires transparency, ethics, and reliability and user testimonials and positive reviews can contribute to building trust (Ediagbonya and Tioluwani 2023).

Data Security Concerns:

- Users may be uncertain to share financial data with new fintech providers due to concerns or misuse and robust data safety measures, regular security audits, are essential to instil confidence and clear privacy policies and effective communication about data protection efforts are vital (Dash et al. 2022).

Transparency (Omarova 2020):

- Fintech companies must maintain transparency in their operations and lack of transparency can erode trust among users who want to understand how their data and finances are managed and open and honest communication can help address this concern.

Addressing these multifaceted challenges and concerns is crucial to ensure that fintech solutions genuinely contribute to financial inclusion while safeguarding user rights and data security. Collaboration among regulators, fintech providers, and civil society is essential.

6.6 Government Initiatives and Policy Support in Promoting Financial Inclusion

Governments across the globe recognize the importance of including the under privileged into the main stream. This section explores the vital role played by governments in promoting financial inclusion as shown in Fig. 6.5.

1. Policy Leadership

Governments set the direction for financial inclusion efforts through policy leadership through national financial inclusion strategies where many governments have developed comprehensive national financial inclusion strategies, outlining their goals, target populations, and action plans. These strategies provide a roadmap for inclusive finance development and financial inclusion policies have the objectives of poverty reduction, assuring economic growth, providing equality, and improved healthcare and education (Sa'ad and Pitchay 2020).

2. Legal and Regulatory Frameworks

Governments often resort to implementation of regulations related to digital identity verification ensuring safe and secure onboarding processes for fintech and regulatory bodies oversee the financial sector and payment systems to maintain stability and protect consumers (Rajeswari and Vijai 2021).

3. Financial Literacy and Education

Financial education may be integrated into school curricula to reach young learners and instill good financial habits early in life. Programmes for adults, especially

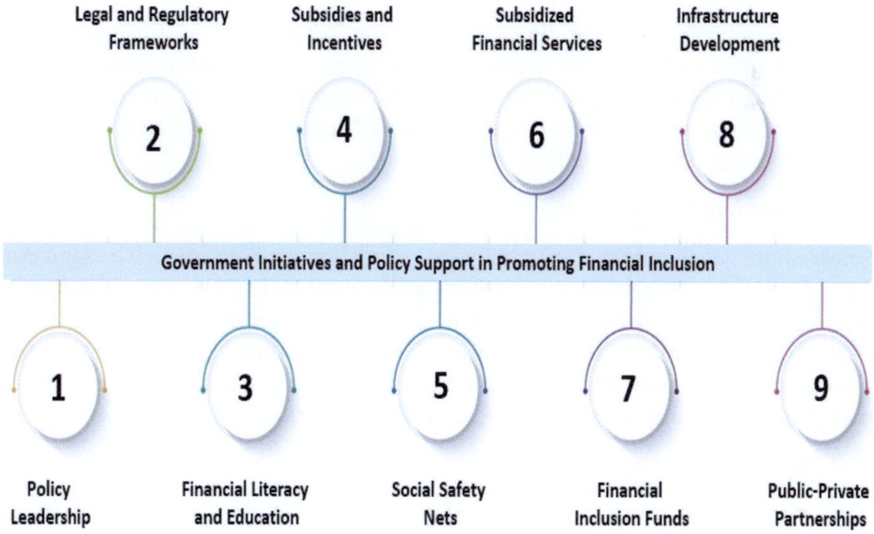

Fig. 6.5 Government initiatives in promoting financial inclusion

in underserved areas, focus on topics like budgeting, saving, investing, and understanding financial products, and given the increasing role of technology in financial services, digital literacy programmes help individuals navigate fintech platforms and digital financial tools (Albarrak and Alokley 2021).

4. Subsidies and Incentives

Governments often provide subsidies or incentives to financial institutions such as Governments may subsidize the cost of digital transactions give tax benefits, such as reduced or waived taxes, can incentivize banks and fintech firms to expand services to underserved areas and may provide financial support or incentives to traditional banks that offer financial products tailored to underserved populations (Muthuswamy and Sharma 2023).

5. Social Safety Nets

Financial inclusion is a means of acting as a social safety net benefits such as governments implement direct benefit transfer (DBT) programs, where the benefits offered can be transferred directly to beneficiaries and governments often may set specific financial inclusion targets for the disbursement of social safety net benefits, ensuring that a significant number of people are included (Jafri 2021).

6. Subsidized Financial Services

Governments may establish microfinance institutions to offer small loans to individuals and small businesses and subsidized insurance products can be offered by governments, such as crop insurance, health insurance, or life insurance and thereby can protect vulnerable segment (Kandpal and Mehrotra 2019; Nagpal 2022).

7. Financial Inclusion Funds

Some governments establish financial inclusion funds such as financial inclusion funds often support fintech startups with promising solutions for expanding access to financial services and funds may provide guarantees to financial institutions or investors, reducing the perceived risks of serving underserved populations (Ediagbonya and Tioluwani 2023).

8. Infrastructure Development

Infrastructure development is critical for extending financial services such as governments may encourage banks to establish branches or agents in underserved regions for better access to banking and expanding mobile network coverage in remote regions enhances the reach of digital financial services and establish digital access centres or community hubs with Internet access allows residents to use fintech solutions and access financial information (Zhykhor et al. 2021; Sharma et al. 2024).

9. Public–Private Partnerships

These are a joint approach for governments to accelerate financial inclusion initiatives and involve collaboration between various stakeholders to pool resources and

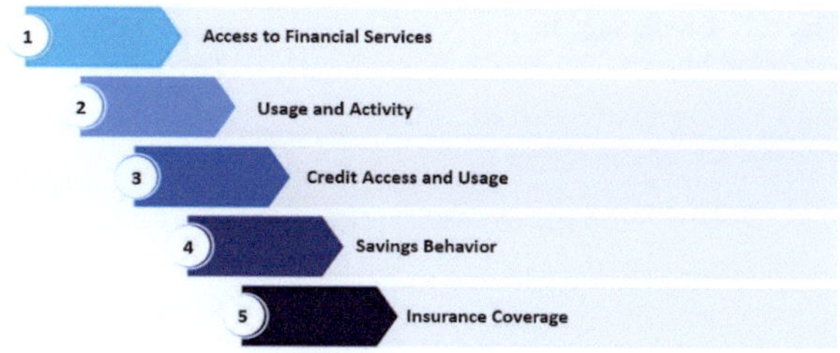

Fig. 6.6 Metrics for evaluating financial inclusion impact

expertise and fintech companies often bring technological innovations to financial inclusion efforts and can assist in technology transfer (Sudarmo et al. 2023).

Impact Assessment and Measurement in Financial Inclusion

With financial inclusion initiatives increasing, measuring the impact of these initiatives is a complex endeavour and some of the metrics used for evaluating financial inclusion impact are as shown in Fig. 6.6.

6.7 Metrics for Evaluating Financial Inclusion Impact

1. Access to Financial Services: The total number of individuals or households with access savings accounts, credit accounts, and insurance policies, etc. and the extent to which under privileged areas are served by various financial services even in remote areas (Yoon et al. 2023).
2. Usage and Activity: The volume and frequency of financial transactions transacted are a metric for assessing the financial inclusion among users and the adoption of digital financial services, which can signify improved reach and usage (Rabbani et al. 2021).
3. Credit Access and Usage: Metrics such as loan penetration such as the percentage of individuals specifically under privileged that have access to credit facilities and loan portfolio size like the total value of loans distributed to previously financially excluded individuals or small enterprises can be measured (Balyuk et al. 2022).
4. Savings Behaviour: Metrics such as savings rates like the percentage of income or financial resources that individuals save in formal financial institutions, indicating their capacity to build financial security and the frequency of Savings such as how often individuals deposit money into their savings accounts, reflecting their commitment to saving regularly (Jünger and Mietzner 2020).
5. Insurance Coverage: Metrics such as insurance penetration and the proportion of the population covered by various insurance products, including life, health, and crop insurance, which provides protection against financial shocks and areas such as claims ratio like the percentage of insurance claims successfully paid out to policyholders, indicating the effectiveness of insurance coverage (Banerjee 2020).

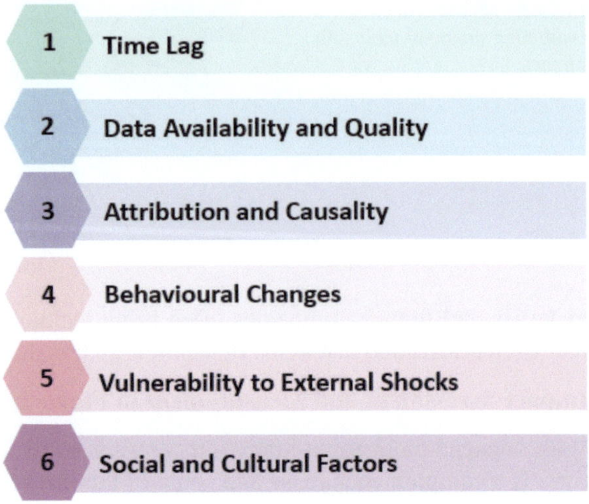

Fig. 6.7 Challenges in assessing impact

Challenges in Assessing Impact (Bazarbash 2019; Moridu 2023; Pambudi et al. 2022)

While these success stories highlight the advantages, several barriers exist in accurately assessing its impact as shown in Fig. 6.7.

1. Time Lag: Measuring the impact of these initiatives on a longer time frame can be challenging. Changes in financial behaviour and economic well-being may take years to manifest fully and thus difficult to assess immediate outcomes accurately.
2. Data Availability and Quality: Reliable data is essential for impact assessment. In many regions, especially underserved areas, data collection and reporting mechanisms may be weak or non-existent, hindering accurate measurement.
3. Attribution and Causality: The question of if financial inclusion directly causes improvements in individuals' lives can be complex and difficult to assess as often various other factors, such as education, healthcare, and economic policies, also influence the livelihood of the people and thus the outcomes, and thus, it becomes difficult.
4. Behavioural Changes: Financial inclusion initiatives often aim to influence financial behaviour, such as saving and borrowing habits. Behavioural changes can be difficult to measure accurately, as they depend on individual choices and circumstances beyond the control of the people.
5. Vulnerability to External Shocks: Economic and environmental shocks, such as financial crises, can disrupt the positive impact of financial inclusion and evaluating resilience to such shocks requires a long-term perspective.
6. Social and Cultural Factors: Social norms and cultural factors play a noteworthy role in financial decisions. Assessing their impact and measuring changes in attitudes and practices is challenging.

Fig. 6.8 Fintech's role in sustainable development goals

Fintech's Role in Sustainable Development Goals (SDGs)

Financial technology, or fintech, has emerged as a powerful force in driving progress towards the United Nations Sustainable Development Goals (SDGs). Fintech plays an important role in aiming to achieve these goals by enhancing financial inclusion, promoting economic growth, and facilitating responsible consumption and production as shown in Fig. 6.8.

1. Goal 1: No Poverty Fintech enables affordable, cost-effective, and fast digital payment options by digital payments and remittances thereby combating poverty and fintech platforms such as Microfinance and P2P Lending can offer microloans and peer-to-peer lending in underserved areas and underserved people and various financial literacy apps of fintech solutions provide financial education thereby empowering people (Zuliansyah et al. 2022).
2. Goal 2: Zero Hunger Fintech products specifically targeted at agricultural finance can provide insurance for small time farmers, thereby mitigating risks that the farmer community face and fintech products targeting supply chain finance can help in improving food supply chains thereby reducing food waste (Mhlanga 2022).
3. Goal 3: Good Health and Well-being Fintech products in the form of telemedicine and health financing can assist telemedicine, making healthcare more accessible for the under privileged and thus assist in their upliftment and

fintech solutions such as health insurance can provide digital insurance products to vulnerable populations, ensuring they have good healthcare services (Farahani et al. 2022).

4. Goal 4: Quality Education: Fintech products in the form of online learning platforms can assist edutech platforms enabling remote and low-cost access to quality education, thereby reducing the education gap and fintech products in the domain of education. Finance can offer education financing, making education more accessible for all (Jamil and Seman 2019).
5. Goal 5: Gender Equality Fintech products targeting financial inclusion for women often aim to promote financial inclusion for women, thereby removing barriers for women prevailing in the society and fintech aimed at digital empowerment can enhance women's digital literacy and empower them to manage their finances independently (Edwards-Dashti 2022).
6. Goal 7: Affordable and Clean Energy: Fintech platforms facilitate investment in renewable energy projects by offering innovative financing models, contributing to the transition to clean energy (Hudaefi 2020).
7. Goal 8: Decent Work and Economic Growth: Fintech platforms offer tools, helping them access capital and manage finances efficiently and can assist in streamline payments for gig workers and freelancers, enhancing economic opportunities (Farahani et al. 2022).
8. Goal 9: Industry, Innovation, and Infrastructure: Fintech builds digital financial infrastructure, expanding access in remote and underserved areas and blockchain technology, a subset of fintech, enhances transparency and traceability in supply chains, promoting sustainable production and trade (Nicanor Franco Riquelme et al. 2020).
9. Goal 10: Reduced Inequality: Fintech narrows the wealth gap by providing previously excluded individuals and communities access to formal financial services and these platforms support asset building and wealth creation through investments and savings opportunities (Demir et al. 2022).
10. Goal 11: Sustainable Cities and Communities: Fintech aids in the growth of smart city technologies, enhancing urban sustainability through efficient transportation, energy management, and waste reduction (Mhlanga 2023).
11. Goal 12: Responsible Consumption and Production: Fintech promotes responsible investments and ethical financial products and fintech-driven supply chain solutions can reduce waste through improved tracking and optimization (Bayram et al. 2022).
12. Goal 13: Climate Action: Fintech supports green finance initiatives by facilitating investments in renewable energy, carbon offset projects, and sustainable agriculture and fintech platforms can streamline the process of investing in green projects by offering user-friendly interfaces and fractional investment options (Sreelekshmi and Biju 2023).
13. Goal 16: Peace, Justice, and Strong Institutions: Fintech can assist in having a transparent Governance in place as it enhances accountability in government operations and can also provide digital identity solutions provided by fintech companies can help establish secure and verifiable identities, reducing fraud and

corruption thereby achieving peace and robust institutions (Pauliukevičienė and Stankevičienė 2022).

14. Goal 17: Partnerships for the Goals: Fintech companies often collaborate with governments, NGOs, and international organizations to advance sustainable development goals thereby facilitating Public–Private Partnerships and also can assist in cross-border payments where fintech improves cross-border payments and international remittances, supporting global partnerships and economic cooperation (Mhlanga 2023).

6.8 Conclusion

Financial inclusion has long been a global challenge and traditional banking systems, with their stringent requirements and physical infrastructure, has often excluded vast segments of the population, particularly in developing economies, whereas the advent of fintech has disrupted this status quo, offering a pathway to financial empowerment for millions. Fintech, with its innovative and technology-driven approach, can leverage digital platforms. Mobile banking allows individuals to open accounts, transfer funds, and do payments through their mobile and digital lending platforms which are extending credit to underserved populations, enabling entrepreneurs to start businesses and individuals to access essential financial support.

Moreover, fintech solutions are addressing the financial needs of specific segments, such as women, youth, and rural populations, and by tailoring products and services to their specific requirements, fintech companies are empowering them, and for example, microfinance platforms are providing small loans to women entrepreneurs, helping them to create livelihoods and contribute to their communities and at the same time are not without its own challenges. Digital literacy, Internet connectivity, and data privacy concerns are some of the obstacles that need to be addressed and collaboration between stakeholders is a must. Fintech has developed as a prevailing tool for bridging the financial divide by offering innovative solutions and reaching underserved populations; fintech is driving economic growth, reducing poverty, and as technology continues to advance, the possibility of fintech to transform lives and create a more inclusive financial ecosystem is immense.

Funding NA.

Conflict of Interest There is no conflict of interest for this research.

References

Albarrak MS, Alokley SA (2021) FinTech: ecosystem, opportunities and challenges in Saudi Arabia. J Risk Fin Manage 14(10):460

Alkhazaleh AMK, Haddad H (2021) How does the Fintech services delivery affect customer satisfaction: a scenario of Jordanian banking sector. Strateg Chang 30(4):405–413

Almuhammadi A (2020) An overview of mobile payments, fintech, and digital wallet in Saudi Arabia. In: 2020 7th International conference on computing for sustainable global development (INDIACom). IEEE, pp 271–278

Anshari M, Almunawar MN, Masri M, Hamdan M, Fithriyah M, Fitri A (2021) Digital wallet in supporting green FinTech sustainability. In 2021 Third international sustainability and resilience conference: climate change. IEEE, pp 352–357

Balyuk T, Berger AN, Hackney J (2022) What is fueling FinTech lending? The role of banking market structure. The role of banking market structure (June 27, 2022)

Banerjee A (2020) Fintech revolution in the gulf countries and Mena region. In: Entrepreneurial finance in emerging markets: exploring tools, techniques, and innovative technologies, pp 335–344

Bayram O, Talay I, Feridun M (2022) Can FinTech promote sustainable finance? Policy lessons from the case of Turkey. Sustainability 14(19):12414

Bazarbash M (2019) Fintech in financial inclusion: machine learning applications in assessing credit risk. International Monetary Fund

Bittini JS, Rambaud SC, Pascual JL, Moro-Visconti R (2022) Business models and sustainability plans in the FinTech, InsurTech, and PropTech industry: evidence from Spain. Sustainability 14(19):12088

Bollaert H, Lopez-de-Silanes F, Schwienbacher A (2021) Fintech and access to finance. J Corp Finan 68:101941

Cambaza E (2023) The role of FinTech in sustainable healthcare development in Sub-Saharan Africa: a narrative review. FinTech 2(3):444–460

Chen Y, Kumara EK, Sivakumar V (2021) Investigation of finance industry on risk awareness model and digital economic growth. Ann Oper Res:1–22

Dash B, Sharma P, Ali A (2022) Federated learning for privacy-preserving: a review of PII data analysis in Fintech. Int J Softw Eng Appl (IJSEA) 13(4)

Datta D (2023) The future of financial inclusion through fintech: a conceptual study in post pandemic India. Sachetas 2(1):11–17

Demir A, Pesqué-Cela V, Altunbas Y, Murinde V (2022) Fintech, financial inclusion and income inequality: a quantile regression approach. Eur J Fin 28(1):86–107

Ediagbonya V, Tioluwani C (2023) The role of fintech in driving financial inclusion in developing and emerging markets: issues, challenges and prospects. Technol Sustain 2(1):100–119

Edwards-Dashti N (2022) FinTech women walk the talk: moving the needle for workplace gender equality in financial services and beyond. Springer Nature

Ezzahid E, Elouaourti Z (2021) Financial inclusion, mobile banking, informal finance and financial exclusion: micro-level evidence from Morocco. Int J Soc Econ 48(7):1060–1086

Farahani MS, Esfahani A, Moghaddam MNF, Ramezani A (2022) The impact of Fintech and artificial intelligence on COVID 19 and sustainable development goals. Int J Innov Manage Econ Soc Sci 2(3):14–31

Gabor D, Brooks S (2020) The digital revolution in financial inclusion: international development in the fintech era. In: Material cultures of financialisation. Routledge, pp 69–82

Ghozi A (2022) The urgency of electronic know your customer (e-kyc): how electronic customer identification works to prevent money laundering in the fintech industry. Diponegoro Law Rev 7(1):34–52

Govindapuram S, Bhupatiraju S, Sirohi RA (2023) Determinants of women's financial inclusion: evidence from India. Ann Public Cooperative Econ 94(1):131–158

Goyal P, Chakrabarti A (2022) Banking the unbanked: the Fintech revolution. In: Intelligent engineering and management for Industry 4.0. Springer International Publishing, Cham, pp 91–105

Harish V, Sharma R (2024) 6 Uncovering the shifting landscape. Green innovations for industrial development and business sustainability: models and implementation strategies, 82

Hoang TG, Nguyen GNT, Le DA (2022) Developments in financial technologies for achieving the Sustainable Development Goals (SDGs): FinTech and SDGs. In: Disruptive technologies and eco-innovation for sustainable development. IGI Global, pp 1–19

Hudaefi FA (2020) How does Islamic fintech promote the SDGs? Qualitative evidence from Indonesia. Qual Res Fin Markets 12(4):353–366

Huibers F (2021) Regulatory response to the rise of fintech credit in The Netherlands. J Risk Fin Manage 14(8):368

Imam T, McInnes A, Colombage S, Grose R (2022) Opportunities and barriers for FinTech in SAARC and ASEAN countries. J Risk Fin Manage 15(2):77

Jafri J (2021) Fintech, philanthropy and development: emerging issues with digital inclusion. Financial Geography Working Paper Series: ISSN 2515−0111, 2

Jameaba MS (2020) Digitization revolution, FinTech disruption, and financial stability: using the case of Indonesian banking ecosystem to highlight wide-ranging digitization opportunities and major challenges (July 16 2, 2020)

Jamil NN, Seman JA (2019) The impact of fintech on the sustainability of Islamic accounting and finance education in Malaysia. J Islamic Social Econ Dev

Joia LA, Cordeiro JPV (2021) Unlocking the potential of fintechs for financial inclusion: a Delphi-based approach. Sustainability 13(21):11675

Jünger M, Mietzner M (2020) Banking goes digital: the adoption of FinTech services by German households. Financ Res Lett 34:101260

Kakinuma Y (2023) Fintech as a gateway to enhance financial literacy. Available at SSRN 4367898

Kandpal V, Mehrotra R (2019) Financial inclusion: the role of fintech and digital financial services in India. Indian J Econ Bus 19(1):85–93

Kukreja G (2021) FinTech adoption in China: challenges, regulations, and opportunities. In: Innovative strategies for implementing FinTech in banking. IGI Global, pp 166–173

Kumari A, Devi NC (2022) The impact of FinTech and blockchain technologies on banking and financial services. Technol Innov Manage Rev 12(1/2)

Liu H, Yao P, Latif S, Aslam S, Iqbal N (2022) Impact of Green financing, FinTech, and financial inclusion on energy efficiency. Environ Sci Poll Res:1–12

Maity S, Sahu TN (2022) Bank branch outreach and access to banking services toward financial inclusion: an experimental evidence. Rajagiri Manage J (ahead-of-print)

Makina D (2019) The potential of FinTech in enabling financial inclusion. In: Extending financial inclusion in Africa. Academic Press, pp 299–318

Mehrban S, Nadeem MW, Hussain M, Ahmed MM, Hakeem O, Saqib S, Khan MA et al (2020) Towards secure FinTech: a survey, taxonomy, and open research challenges. IEEE Access 8:23391–23406

Mehrotra A (2019) Financial inclusion through fintech–a case of lost focus. In: 2019 International conference on automation, computational and technology management (ICACTM). IEEE, pp 103–107

Mehrotra A, Menon S (2021) Second round of FinTech-trends and challenges. In: 2021 2nd International conference on computation, automation and knowledge management (ICCAKM). IEEE, pp 243–248

Mhlanga D (2022) Prospects and challenges of digital financial inclusion/fintech innovation in the fourth industrial revolution. In: Digital financial inclusion: revisiting poverty theories in the context of the fourth industrial revolution, pp 163–182

Mhlanga D (2023) FinTech and artificial intelligence for sustainable development. In: Sustainable development goals series. https://doi.org/10.1007/978-3-031-37776-1

Moridu I (2023) The impact of financial statement quality on investment decision making: a descriptive study of the banking sector in West Java. ES Acc Fin 1(03):169–175

Muthuswamy VV, Sharma A (2023) Role of emerging financial technology on environmental and social governance of textile companies in Saudi Arabia. Cuadernos De Economía 46(130):64–72

Nagpal P (2022) Online business issues and strategies to overcome it–indian perceptive. SJCC Manag Res Rev 1–10. https://doi.org/10.35737/sjccmrr/v12/i1/2022/151

Naz F, Karim S, Houcine A, Naeem MA (2022) Fintech growth during COVID-19 in MENA region: current challenges and future prospects. Electron Comm Res:1–22

Nelaturu K, Du H, Le DP (2022) A review of blockchain in fintech: taxonomy, challenges, and future directions. Cryptography 6(2):18

Nicanor Franco Riquelme J, Lemus Aguilar I, Ordieres Meré J, Bello Garcia A (2020) SDGs & innovation: the discourse of FinTech companies through social media analysis. Acad Manage Glob Proc 183

Odei-Appiah S, Wiredu G, Adjei JK (2022) Fintech use, digital divide and financial inclusion. Digit Policy, Regulation Govern 24(5):435–448

Omarova ST (2020) Technology v technocracy: Fintech as a regulatory challenge. J Fin Regulation 6(1):75–124

Ozili PK (2020) Financial inclusion and Fintech during COVID-19 crisis: policy solutions. Company Lawyer J 8

Pambudi PDL, Nurrosyidah A, Justitia A, Putra CD (2022) A proposed assessment tool for Sharia-compliant equity crowdfunding projects utilizing TOPSIS. Indonesian Scholars Scient Summit Taiwan Proc 4:49–57

Panos GA, Wilson JO (2020) Financial literacy and responsible finance in the FinTech era: capabilities and challenges. Eur J Fin 26(4–5):297–301

Pauliukevičienė G, Stankevičienė J (2022) Assessment of the impact of sustainable development goals indicators on the sustainable development of FinTech Industry. Bus Manage:290–298

Prakash C, Saini R, Sharma R (2021) Role of internet of things (IoT) in sustaining disruptive businesses. Internet of things and businesses in a disruptive economy

Putritama A (2019) The mobile payment fintech continuance usage intention in Indonesia. Jurnal Economia 15(2):243–258

Rabbani MR (2022) Fintech innovations, scope, challenges, and implications in Islamic Finance: a systematic analysis. Int J Comput Digit Syst 11(1):1–28

Rabbani AP, Alamsyah A, Widiyanesti S (2021) An effort to measure customer relationship performance in Indonesia's Fintech industry. arXiv preprint arXiv:2102.08262

Rajeswari P, Vijai C (2021) Fintech industry in India: the revolutionized finance sector. Eur J Mol Clin Med 8(11):4300–4306

Rauniyar K, Rauniyar K, Sah DK (2021) Role of fintech and innovations for improvising digital financial inclusion. Int J Innov Sci Res Technol 6(5):1419–1424

Rhanoui S (2023) Enhancing financial inclusion using Fintech: development scenario for the bank card. Int Bus Res 16(5):1–38

Rusadi FARP, Benuf K (2020) Fintech peer to peer lending as a financing alternative for the development MSMEs in Indonesia. Legality: Jurnal Ilmiah Hukum 28(2):232–244

Sa'ad AA, Pitchay ABA (2020) Financing SMEs through the Islamic qard al-Hasan: a proposal of Islamic Fintech interest-free lending for government initiatives. In: Impact of financial technology (FinTech) on Islamic finance and financial stability. IGI Global, pp 54–64

Sampat B, Mogaji E, Nguyen NP (2023) The dark side of FinTech in financial services: a qualitative enquiry into FinTech developers' perspective. Int J Bank Mark

Senyo PK, Osabutey EL (2020) Unearthing antecedents to financial inclusion through FinTech innovations. Technovation 98:102155

Sharma R, Harish V, Rana G (2024) Navigating Risk In The Age Of Artificial Intelligence: Assessing And Identifying Risks With AI Strategies. Educational Administration: Theory and Practice 30(4):2130–2140

Sreelekshmi G, Biju AV (2023) Exploring the interconnection of FinTech and climate sustainability: scoping through a qualitative methodology. In: The sustainable Fintech revolution: building a greener future for finance. IGI Global, pp 292–311

Sudarmo S, Mursalin A, Widyaningrum M, Khasanah K, Utama NU (2023) Bridging the fintech gender gap: strategies for enhancing women's access to financial services. Int J Human Social Sci Bus (INJOSS) 2(3):482–504

Sunardi R, Hamidah H, Buchdadi AD, Purwana D (2022) Factors determining adoption of fintech peer-to-peer lending platform: an empirical study in Indonesia. J Asian Fin Econ Bus 9(1):43–51

Yoon SS, Lee H, Oh I (2023) Differential impact of Fintech and GDP on bank performance: global evidence. J Risk Fin Manage 16(7):304

Yuniarti S, Rasyid A (2020) Consumer protection in lending Fintech transaction in Indonesia: opportunities and challenges. J Phys: Conf Ser 1477(5):052016

Zhykhor O, Ryeznikov V, Pohribna N, Delas V (2021) Digital finance and fintech applications in Ukraine: current status and development prospects. In: Conference on integrated computer technologies in mechanical engineering–synergetic engineering. Springer International Publishing, Cham, pp 979–991

Zuliansyah A, Pratomo D, Supriyaningsih O (2022) The role of financial technology (Fintech) in ZIS management to overcome poverty. Indonesian Interdiscip J Sharia Econ (IIJSE) 5(1):203–224

Chapter 7
Corporate Climate Risk Disclosure

C. A. Pinky Agarwal, Neeraj Gupta, and Debapriya Samal

Abstract The last few decades have witnessed a number of climate movements and advocacy campaigns aimed at connecting climate change with decision-making in various sectors of society including businesses. As a result, climate risk disclosure has emerged as a necessary element of corporate governance and financial reporting alongside the development of policies and goodwill action toward climate risks. Climate risk disclosure is essential for all stakeholders including investors, regulators, consumers, and society to provide information as to how owners of the companies are dealing with climate change-related risks. Such disclosure is critical for Improving transparency, developing accountability, and showing commitment to address climate change issues. The tools from the Task Force on Climate-related Financial Disclosures (TCFD), Global Reporting Initiative (GRI), and Sustainability Accounting Standards Board (SASB) are being developed in response to the demand for comprehensive climate risk disclosure guidelines. These frameworks assist companies in reporting climate risks alongside risk management mechanisms. Also, industry-specific solutions highlight the particular climate threat that various sectors encounter, thereby underscoring the significance of customized disclosure guidelines. The landscape of climate risk disclosure is changing because of technological progress including big data, artificial intelligence, and blockchain since they improve data analysis, security, and reporting capabilities. However, such developments do not eliminate the challenges of data availability, data availability, standardization, and transparency versus competition. To improve corporate climate risk reporting, it is recommended that companies follow industry standards, use resources and reporting

C. A. P. Agarwal (✉)
Associate Professor, Management-Finance, Atlas SkillTech University, Mumbai, India
e-mail: pinkyagarwalca@gmail.com

N. Gupta
Associate Professor, PGDM-Finance, NL Dalmia Institute of Management Studies & Research, Raigarh, MH, India
e-mail: neeraj.gupta@nldalmia.edu.in

D. Samal
Assistant Professor, PGDM-Finance, NL Dalmia Institute of Management Studies & Research, Raigarh, MH, India
e-mail: debapriya.samal@gmail.com

R. Kumar (ed.), *Sustainable Finance*, Sustainability Solutions,
https://doi.org/10.1007/978-3-031-97962-0_7

technologies, focus on particular sectors, strengthen stakeholder communications and enhance the quality of the assurance and verification processes. Better disclosures further sustainability objectives, enhance policy making, and provide for long term sustainability by embedding climate risks in the corporate strategy and risk management. The development of climate risk reporting stresses its relevance to the corporate strategy and governance of the firms. In properly implemented standardization, technological solutions, and timely active interactions with stakeholders' companies will improve the climate risk disclosure practice for the good of policy realization and for further sustainability goals.

Keywords Sustainability · Corporate risk · Technologies · Climate change · Sustainability reporting

7.1 Introduction to Corporate Climate Risk Disclosure

The climate crisis is now often viewed as a business crisis, and many companies have begun to integrate climate risks into their overall strategic approach—one that could provide a competitive advantage. Climate risk has financial impacts—severe and various impacts are felt through almost every aspect of the business and its value chain. Such physical risks of climate change such as extreme weather events and their impact on infrastructure as well as supply chain disruptions can have significant costs (Hope and Laasch 2024). Other risks such as transition risks from changes in policy or carbon pricing or shifts in the market demand similarly threaten the profitability or market valuation. In addition, a smart management of climate risk addresses these challenges while promoting sustainability by suggesting active measures—including lowering emissions of greenhouse gases, improving energy efficiency, or switching to renewable sources of energy—which organizations can pursue as part of the overall climate strategy (Marković et al. 2016). Such measures not only avoid risk, but also bring in new opportunities and cost savings which translates to improvement of the company's economic performance, leading to investment interests, and advancing how the company is perceived in the market. In embedding climate risk within their overall corporate strategy, companies can also find themselves contributing to the higher-level global sustainability targets including within the Paris Agreement or the UN Sustainable Development Goals (SDGs).

For the entire spectrum of evaluation and management of climate-related risks, organizations engage in scenario analysis and stress tests along with quantitative and qualitative assessments. Scenario analysis includes preparing and looking into different climate scenarios in order to assess the damage they could do. Stress tests measure the capacity to remain financially viable during extreme climatic conditions. Quantitative assessments utilize statistical means to assess the climate risks, while qualitative assessments depend on expert views and storytelling (Ben-Ama et al. 2023).

7.2 Objectives of Climate Risk Disclosure

Disclosure of climate-related risk has emerged as an essential feature of corporate governance in the sense that the level and potential scope of corporate exposure to climate-related risk can and should be disclosed, and what the corporate method is to address that risk can and should be disclosed (Liu et al. 2024). This disclosure outlines several key objectives, such as stakeholder response to the demands and priorities of stakeholders, and business agility and resilience.

The three primary goals of climate risk disclosure are:

- **Enhancing Transparency and Accountability**: When risks climate is disclosed, companies make their stakeholders aware of how they are committed toward climate impacts reduction and targets such as net zero emissions. Stakeholders are able to track the performance of the company by setting goals such as greenhouse gas reductions, energy intensity improvements, or investing in climate adapted infrastructure. Investors, regulators, and consumers become more trustworthy which further reinforces the company policies advocating for sustainability. It also increases resilience as it encourages organizations to evaluate their weaknesses and adopt effective strategies on climate risks for the prevention of potential disruptions.
- **Meeting Stakeholder Expectations**: There is also such workers not only but increasingly also ones who will be employed by companies also striving for sustainability. Carbon risk disclosure aligns with who it may matter that their manager is taking any action in relation to climate change, an extremely important consideration to human resources selection and retention. Investors and internal stakeholders, in the form of the board of directors and senior management, are requesting enhanced climate disclosures including, among others, the company's risk and business strategy. This data is of significant relevance for decisions regarding long term resource distribution and strategy, allowing one to coordinate with investors whose investment approach is based on sustainable and responsible investment.
- **Addressing Customer Expectations and Regulatory Compliance**: Today's consumers are sensitive to environmental calls for consequences and are seeking more information about how, and at what potential cost, companies are acting—and not otherwise acting, in the areas related to climate change challenges. The disclosure of climate permits companies to attest to their commitment to sustainability and as a result, increase the brand loyalty and appeal both to sustainability consumers and environmentalists. However, the information on the delivery side may be misleading if it seems to be a reaction to climate change but actually be an intentional effort to change and destroy it for the sake of the customers and the company. Furthermore, meeting the regulatory reporting obligations for climate risk, at least in the international regulatory architecture, helps to maintain system-wide financial stability withstand demands from climate-related shocks.

Compliance not only assures a regulatory compliance but also it is an environmentally responsible undertaking, a matter to be expected to become, in future, a competitive advantage.

7.3 Integrating Climate Risk into Corporate Strategy

Climate risk reduction demands effective governance frameworks and competent risk management processes, as well as business model grounded on the sustainability of rethinking. It is possible to achieve this organizational by targeting the climate risk within the strategic work activities to implement, thus developing the competence in terms of preparedness and resilience to the consequences of the climate change and the satisfaction of the actors who take part in the work (PwC 2022).

7.3.1 Reporting and Disclosure Practices

Effective reporting and disclosure practices are essential for communicating an organization's approach to climate risk management.

7.3.1.1 Key Components of Climate Risk Reports

- **Governance**: Reports should detail the board's oversight role and the responsibilities of senior management in climate risk management.
- **Strategy**: Disclosures must explain how companies incorporate climate risks into overall business strategy and planning.
- **Risk Management**: Companies should outline processes for identifying, assessing, and managing climate-related risks.
- **Metrics and Targets**: Reporting must include quantitative data on GHG emissions, energy consumption, and progress toward targets Yoo 2023).

7.3.2 Governance Structures in Climate Risk Management

Good governance structures are essential for organizations to efficiently respond to climate risks. Such structures enable climate-related aspects to be incorporated and dealt with in decision-making processes at the highest level of the corporation.

- **Roles of Boards and Executive Teams in Climate Risk Management**

Board of Directors

The board of directors plays a vital role in overseeing climate risk management (Liu et al. 2024). Their key responsibilities include:

- **Setting Climate Risk Strategy**: The board establishes the strategic direction for climate risk management, ensuring alignment with the organization's long-term objectives. This involves the approval of policies related to climate risks.
- **Oversight and Accountability**: The board monitors the implementation of climate strategies and holds the executive team accountable for their performance. They regularly review reports detailing the company's climate risk exposure and mitigation efforts.
- **Stakeholder Engagement**: The board interacts with stakeholders—such as investors, regulators, and non-governmental organizations—to grasp their concerns regarding climate risks and align the company's climate strategy with external expectations.

Executive Teams

The executive team, including the CEO, CFO, and senior executives, is responsible for the operational aspects of climate risk management. Their roles encompass:

- **Implementation of Strategy**: Executives are tasked with executing the board's climate risk strategy by incorporating climate considerations into daily operations and decision-making processes.
- **Resource Allocation**: They allocate resources for managing climate risks, including investments in resilience measures and technologies.
- **Reporting and Communication**: Executives ensure timely and accurate reporting on climate risks and management efforts to both the board and external stakeholders.

Many organizations establish dedicated committees, such as a Climate Risk or Sustainability Committee, to focus specifically on climate-related issues. These committees are responsible for:

- **Monitoring and Reporting**: They review climate risk data, assess progress against goals, and make recommendations to the board.
- **Policy Development**: Committees develop and update climate risk policies and procedures, ensuring compliance with current regulations and best practices.
- **Engagement and Advocacy**: They engage with external stakeholders and advocate for climate-related initiatives and policies.

7.3.3 Strategic Planning

Strategic planning involves aligning business strategies with climate resilience and sustainability goals for long-term success.

- **Aligning Business Strategies with Climate Resilience and Sustainability Goals**
  - **Integration into Strategic Objectives**: Businesses must integrate climate resilience and sustainability goals into the company's overall strategic goals and also set goals with respect to greenhouse gas emissions and energy efficiency.
  - **Long-Term Planning**: Developing strategies that account for the impacts of climate change on business operations is essential. This involves:

    Scenario Planning: Integrating climate scenarios into long-term plans to prepare for future changes.
    Innovation and R&D: Investing in research to drive innovation in sustainable products and processes.

7.3.4 Risk Management Processes

Risk management processes are essential for identifying, assessing, and mitigating climate risks, enabling organizations to understand their exposure and implement strategies effectively.

7.3.4.1 Identifying Climate Risks

- **Risk Identification**: The initial step is to identified the climate-related risks which could affect the business, i.e., physical risks (e.g., extreme events, long-term climate dynamics) and transition risks (e.g., regulatory/technological change) (Kumar et al. 2023a; b, c, d, e).

  Common methods include:

  Risk Assessments: Conducting assessments to pinpoint vulnerabilities and potential impacts on operations and supply chains.
  Stakeholder Input: Gathering feedback from stakeholders to identify emerging risks and concerns.

7.3.4.2 Assessing Climate Risks

- **Risk Assessment Frameworks**: Organizations use various frameworks to evaluate the significance and likelihood of identified climate risks, including:
  - **Qualitative Assessments**: Utilizing expert judgment for risk evaluation.
  - **Quantitative Assessments**: Applying numerical data and models to estimate potential financial impacts from climate risks.

  - **Scenario Analysis**: Evaluating risks under different climate scenarios to understand potential future impacts.
- **Prioritization**: After assessing risks, companies prioritize them based on their potential impact and likelihood, focusing resources on the most critical areas.

7.3.4.3 Mitigating Climate Risks

- **Risk Mitigation Strategies**: Developing and implementing strategies to address climate risks is crucial. This includes:
  - **Operational Changes**: Modifying practices to minimize exposure, such as upgrading infrastructure or diversifying supply chains.
  - **Investment in Resilience**: Allocating funds for technologies that enhance climate resilience, such as renewable energy sources.
  - **Insurance and Financial Instruments**: Using insurance and other financial instruments to manage risks associated with climate events.

7.3.5 Performance Metrics and Reporting

Establishing metrics and key performance indicators (KPIs) to track progress toward climate goals allows for effective performance evaluation.

- **Engaging with Stakeholders**
  - **Stakeholder Engagement**: Involving stakeholders to understand their expectations and incorporating their feedback into planning is vital. This includes:

    Investor Relations: Communicating climate strategies and performance to foster trust among investors.
    Customer and Supplier Engagement: Collaborating with customers and suppliers on sustainable practices.
- **Policy Advocacy**: Engaging in policy discussions and advocating for supportive regulations enhances alignment with broader climate goals.

7.4 Regulatory Frameworks for Climate Disclosure

Reporting on climate also dramatically changed the rules, which have been shaped by a series of global standards that restrict the size of reporting to be both understandable and persuasive for readers, as well as standardized in the disciplines. Some of these frameworks are the Task Force on Climate-related Financial Disclosures (TCFD), Global Reporting Initiative (GRI) and the Sustainability Accounting Standards Board (SASB) (Aleksei 2024).

- **Task Force on Climate-Related Financial Disclosures (TCFD)**: The TCFD was established by the Financial Stability Board in 2015 and was chaired by Mark Carney and Michael Bloomberg, with the goal of including climate disclosures in consistent and voluntary reporting which would provide an opportunity for organizations to inform investors, insurers and the wider society about climate related risks and opportunities. The TCFD framework consists of four core components:
  - **Governance**: Details organizational oversight concerning climate risks, including roles for the board and management.
  - **Strategy**: Examines potential impacts of climate risks on business strategy and financial planning.
  - **Risk Management**: Outlines how climate risks are identified, assessed, and managed within the overall risk framework.
  - **Metrics and Targets**: Specifies metrics for evaluating climate risks and tracking progress.

The regulatory framework concerning climate risk reporting has been significantly impacted by the recommendations of the TCFD, with several jurisdictions supporting or even requiring TCFD aligned disclosures. Its previous guidance has received support worldwide from governments, financial institutions, and corporations, with the UK, New Zealand, and Japan making it compulsory or nearly so (Ngo et al. 2022). The TCFD has established itself as a leading global framework and has incorporated reporting standards of others and supported the evolution of climate-related financial disclosures. Businesses that apply the recommendations of TCFD are in a better position to address stakeholder's needs, build resilience, and ensure operational sustainability in the long term, thereby embedding climate risk disclosure in the contemporary business environment.

- **Global Reporting Initiative (GRI)**: The GRI was constituted back in 1997 and now stands a leading framework in sustainability reporting that enables organizations to report on their economic, environmental and societal impacts including their impacts on climate. Companies need to even acknowledge and report on significant sustainability issues including climate change, human rights, corruption and resource use and that's also the core mission of the GRI. In order to support sustainable development decision making, the GRI advances the cause of transparency and accountability.

The GRI Standards employ a modular structure that facilitates extensive reporting across various sustainability topics, categorized into three main series:

Universal Standards: These guidelines cover all organizations and set out specific objectives and principles related to reporting and management, as well as, general disclosures regarding the governance and ethical principles and stakeholder relations management for that organization.

Topic-Specific Standards: They address certain components of sustainability such as: economic performance, the environment, labor, and human rights as well as anti-corruption components. Relevant to climate risk standards are also the

GHG emissions (GRI 305) and energy (GRI 302) and water use standards (GRI 303).

Sector Standards: These standards are developed with respect to the specific industries faced. For example, the financial services sector may have additional standards that deal with the risks of climate change of investment and lending.

The GRI Standards are widely adopted by companies all over the world and consequently represent the most widespread reporting platform for sustainability. Corporations are providing granular data on the environmental footprint of companies to investors, to customers, and to regulators with respect to climate-related risk, in accordance with GRI standards. The GRI has played a key role in defining sustainability reporting standards and outlining companion frameworks (e.g., TCFD and SASB) and is highly regarded for its approach to materiality and stakeholder involvement.

- **Sustainability Accounting Standards Board (SASB)**: SASB (established in 2011) is advocating industry-dedicated standards for the dissemination of finance-relevant sustainability disclosures. Its overarching aim evidently is to inform investors and financial partners of the sustainability-based opportunities and risks-which may impact on how a given company operates (IFRS, 2024). SASB standards are being designed to be incorporated in the usual financial disclosures of companies as a shared format that will convey sustainability and financial information and help corporations' investors to understand the effect of sustainability challenges (e.g., climate change) on a company's value and risk.

Moreover, SASB standards are industry based as they recognize that climate risks and sustainability issues can be sector specific. SASB has developed standards for 77 industries across 11 sectors, which are designed to the specific risks and opportunities each of these sectors possess.

The key components of the SASB standards include the following:

Materiality: The standards highlight sustainability issues most likely to influence financial performance within specific industries, safeguarding that companies report on the most appropriate factors and provide decision-useful information to investors (Ennew et al. 2024).

Financial Impact: The financial implications of sustainability issues highlighted by SASB standard, including climate related risk, protecting various aspects such as costs, assets, revenue and capital expenditures.

Quantitative and Qualitative Metrics: Quantitative and qualitative indicators of sustainability reporting are employed as criteria. SASB quantifies climate risk with metrics of greenhouse gas emissions, energy use, water use, and climate change regulatory risk.

Sustainability Accounting Standards Board (SASB) standards are becoming more popularly used by investors–largely in the US–as a supplement to traditional financial reports. Increasing use of the SASB standards in sustainability reporting is becoming a "fact of life", not just in the dialogue, but also for companies, particularly those in climate impacted industries such as energy and transport. In collaboration with the International Integrated Reporting Council (IIRC), SASB to Value Reporting Foundation restructured alignment and globalizing financial and sustainability reporting (Norton et al. 2024).

- **European Union's Non-Financial Reporting Directive (NFRD)**: Non-financial reporting directive (NFRD), approved by the EU in 2014, is in particular designed for enhancing reporting of non-financial performance (NFRD). Hence, disclosure is also demanded in respect of such entities at public interest level, i.e., public listed companies and banks having more than 500 employees required to disclose public interest data of public interest data of non-financial and diversity data. This Directive has the purpose of disclosing nonfinancial performance of specific companies, with particular emphasis on the environmental theme and on the performance based on the environmental stand and on the management of the sustainability-related risk. The key objectives of the NFRD include:
  - **Enhancing Transparency**: Improving the quality and comparability of non-financial information to better assess ESG performance.
  - **Promoting Sustainability**: Encouraging companies to consider long-term sustainability in their operations and decision-making.
  - **Facilitating Investment Decisions**: Providing investors with essential information for evaluating sustainability risks and opportunities.

  The key reporting requirements include:

- **Environmental Matters**: Details on environmental impacts, GHG emissions, and resource use.
- **Social and Employee-Related Matters**: Information on employee treatment, working conditions, and community (Flammer 2023)

- **The U.S. Securities and Exchange Commission (SEC)** in order to emphasize on climate related disclosures, issued guidelines for listed companies in 2010. The guidelines outlined the climate related disclosure requirements regarding the impacts of climate change, its impact on the regulatory requirements and the international agreements. To strengthen transparency and consistency in the disclosures, SEC gathered inputs from the general public in 2021. This lead to the introduction of new rules pertaining to the governance, operations and greenhouse emissions which were aligned with the earlier rules like TCFD etc. (Brown and Wald 2023).
- **The Business Responsibility and Sustainability Reporting (BRSR)** framework in India, which aligns with the National Guidelines on Responsible Business Conduct (NGRBCs), was instituted by the Securities and Exchange Board of India (SEBI) in 2021. It required 1000 leading firms to reveal information regarding environmental, social, and governance (ESG) issues (Mohanty et al. 2024). The basic objective of the framework is to promote transparency, accountability, and

ethical practices, transparent, and accountable business practices, aligning. BRSR requires disclosures under three main sections:

- **General Disclosures**: Requires company details, products/services, operations, employees, etc.
- **Management and Process Disclosures**: Reflects adherence to NGRBC principles through company policies and processes.
- **Principle-wise Performance Disclosures**: Ratifies the company's performance in complying with NGRBC principles through key processes and decisions.

With the increase in the global awareness toward climate change, convergence toward climate risk disclosure practices is being witnessed throughout the world. This necessitates the harmonization of the universal disclosure standards across the jurisdictions. Though the goal of the frameworks for climate risk disclosure is the same across jurisdictions, there are variations in breadth, statutory requirements and extent of the solutions. The EU's NFRD (and the increasingly amended CSRD) combination is perhaps the most comprehensive and stringent regulatory landscape requiring qualitative, comparable climate reporting. On the other hand, the US SEC has long promoted materiality but has increasingly been moving toward more prescriptive disclosure requirements. International best practice (e.g., TCFD) provides a principled, general, and open-ended, consensus-based, solution, although despite unprecedented community-level enthusiasm, there is no lack of such mechanisms (Kumar et al. 2024a). At the same time as the increasing visibility of the risks and challenges that climate change brings to countries all over the world, there is also an increasing convergence in their practices for disclosing climate change risk (Kumar et al. 2024b). Businesses operating across jurisdiction will have to manage this complex regulatory framework in their efforts to meet regulatory requirements, investor requirements, and stakeholder expectations while prudently managing and disclosing climate-related risk (Venturelli et al. 2024).

7.4.1 Industry-Specific Risks Disclosure and Practice

Tailoring climate risk disclosures to different sectors is essential due to the unique challenges and risks associated with climate change.

Below are the specific practices relevant to various industries:

- **Energy Sector**: The risk of impact from climate change in fossil fuel-based companies due to regulations and policies, availability of technology, and changes in market trends toward renewable energy has become critical concern in energy industry firms. Most disclosures emphasize transition risks pertaining to changes within the market structure with regard to pricing carbon credits or emissions, or making the investments to implement sustainable technologies. Similarly, other risks including physical risks that is the destruction or damage to energy systems from extreme weather are pertinent here as well (Zhang et al. 2024). It is important that these disclosures address how these risks will be managed and how the company plans to adapt its business in a transition to a clean energy era.

- **Finance Sector**: Financial institutions confront both transition and physical risks. Disclosures related to climate are predominantly on the impact of climate risk on investment portfolios, investment strategies, and risk architecture. This field is in particular interest in the economic impact of climate risks to investments such as asset disbalance and rises in account defaults. Successful disclosures for finance are the incorporation of climate risk into investment practices, and incorporating sustainability into clients' relationships.
- **Manufacturing Sector**: "The manufacturing sector's disclosures typically address both transition and physical risks related to the impacts of climate change on supply chains, production processes, and regulatory compliance. Companies must report on the risks associated with supply chain disruptions due to extreme weather events, energy consumption and efficiency, and regulatory compliance costs. Disclosures may include assessments of the resilience of supply chains to climate-related disruptions, strategies for reducing carbon footprints, and investments in technologies to improve energy efficiency and reduce emissions."

7.5 Compliance and Enforcement

- **Mechanisms for Ensuring Adherence**: There are regulatory authorities that supervise adherence to the climate specific disclosure requirements and provide instruction (Lei and Zhang 2024). Certain disclosures may be subject to an audit requirement or included in annual reports and specific submission platforms may be mandated.
- **Enforcement Actions**:
  - **Penalties and Fines**: Non-compliance can lead to financial penalties and legal actions.
  - **Public Disclosure**: Instances of non-compliance may be reported publicly, damaging a company's reputation.

7.6 Impact of Climate Risk Disclosure on Financial Performance

Stakeholders emphasizing sustainability have forced the need for climate risk appraisal to be tackled with urgency. Such climate disclosures have an impact on the perception of the investors about the business, the cost of capital and the valuation of that business in the market.

7.6.1 *Investor Perceptions and Behavior*

- **Influence on Investment Decisions**: Higher equity risk reduction companies that disclose the risk factors lead to better post investment in the shareholders' perception (Ilhan et al. 2023).
- **Investor Confidence**: More confident investors willing to invest can result from a more comprehensive disclosure.
- **Investment Preferences**: Informatively robust disclosures are not disadvantageous because the there is a growing trend among most investors to go green.

7.6.2 *Cost of Capital*

- **Relationship Between Transparency and Access to Financing**: Disclosures that are better in clarity would raise cheaper amounts of debts and equity funds.
- **Access to Green Finance**: Access to issuance of green bonds and other eco-friendly sources of finances may be made easier by appropriate handling of climate change risks (Wu et al. 2024).

7.6.3 *Market Valuation*

- **Effects of Climate Risk Disclosure on Company Valuation**: Strong companies have better chances of being given higher market values as the investors of such companies do care about their value in the long run (Vestrelli et al. 2024).
- **Impact on Shareholder Value**: Defensive proactive disclosures management creates anticipation around long term value creation.

7.7 Best Practices in Reporting

- **Ensuring Clarity, Consistency, and Comparability**: Reports must be written in a manner that depicts clarity and consistency which is measured using the relevant indicators over a specified period (Day et al. 2022).
- **Transparency and Detail**: Thorough disclosures create confidence among stakeholders as they give detailed explanations about the specific problem covered and the management techniques employed.
- **Stakeholder Engagement**: Involving stakeholders in the reporting process to identify the information required strengthens the reporting process.
- **Continuous Improvement**: Updating and revising the practices of reporting frequently makes them applicable and efficient.

7.8 Case Studies and Best Practices in Climate Risk Disclosures

7.8.1 Leading Companies

Several organizations exemplify exemplary climate risk disclosure practices by utilizing established frameworks to convey their strategies and progress transparently:

- **Unilever**: The TCFD framework, as adopted, is widely appreciated for the coverage and depth of climate risk disclosures it implements, and Unilever harnesses it to deliver an open account of climate risk and opportunity. They are succinct and tightly argued, with very deft and impeccable integration of climate-related material with the rest of the business plan. Unilever performed deep scenario analysis not only assessing physical risks, but also transitioning risks and also monitoring the trajectory to sustainability targets through metrics tracked openly in line with the Global Reporting Initiative (GRI) principles, for further improvement of transparency.
- **Microsoft**: Microsoft's climate risk disclosures were the first in the technology industry. The company is going to be carbon negative by 2030 which it also mentions in its annual sustainability report with relevant strategies and progress. Microsoft practices scenario making in order usually to estimate possible climate change effects over operations and supply chains focusing on carbon foot prints and energy consumption. It is worth mentioning that TCFD recommendations are observed by them while reporting in a detailed and transparent manner.
- **Tesla**: When we talk of climate risk integration within business strategy, we can consider Tesla as a case within the auto industry. The firm discloses risks related to regulatory shifts, climate change trends and their relation to non-automotive based technology, and market competition. Tesla focuses on emission reduction with electric vehicles and other sustainable energy development in its metrics and targets, which are compliant with TCFD and SASB standards.

7.8.2 Sector-Specific

Best practices differ significantly across sectors:

- **Energy Sector**:
  - **ExxonMobil**: Company climate risk disclosures are broad ranging including physical and transition risks which is concentrated on regulatory and market forces in the oil and gas sector. ExxonMobil incorporates scenario analysis into the development of future scenarios and reports on emissions, clean energy efforts, and policies and practices related to adaptation to physical climate risks.

- **Finance Sector**:
  - **Goldman Sachs**: Goldman Sachs provides comprehensive disclosures on how climate risks affect its investment portfolios and financial performance. The company uses the TCFD framework to report on governance structures, risk management processes, and climate-related financial metrics. Best practices in the finance sector involve integrating climate risk assessments into investment decision-making, reporting on climate-related exposure, and engaging with clients on sustainability issues.
- **Manufacturing Sector**:
  - **General Electric (GE)**: In the case of GE, climate risk disclosures illustrate the company's initiative to reduce emissions and to improve energy efficiency of all their manufacturing plants. The company also states it is pursuing such a direction in order to reach sustainability goals. For instance, it indicates it is investing on reduction of greenhouse gases' emission and green technologies' development. Good mechanical manufacturing best practices are well established and serve to guide reporting of emission reductions, energy consumption, and material use, as well as to facilitate the contextualization of innovation in support of sustainable development.

The case studies of leading companies and learning from both successful and unsuccessful disclosure efforts provide valuable insights into effective climate risk disclosure practices. Insights from successful and unsuccessful disclosure efforts reveal critical elements of effective climate risk disclosure practices:

- **Successful Disclosure Efforts**:
  - **Clarity and Specificity**: Effective disclosures provide precise and actionable information, building trust with stakeholders through solid data and metrics.
  - **Alignment with Frameworks**: Companies adhering to established reporting frameworks, such as TCFD, GRI, and SASB, ensure comprehensive, consistent, and comparable disclosures that enhance credibility.
  - **Integration into Business Strategy**: Disclosures that are embedded within broader business strategies gain relevance and impact, illustrating how climate risks and opportunities inform strategic planning.
- **Unsuccessful Disclosure Efforts**:
  - **Lack of Detail and Transparency**: Vague disclosures can diminish credibility and fail to address stakeholder concerns. Incomplete information hinders trust and demonstrates ineffective climate risk management.
  - **Inconsistency and Misalignment**: Discrepancies in disclosures, particularly when recognized frameworks are not followed, lead to confusion and hinder effective stakeholder engagement.

- **Inadequate Risk Assessment**: Disclosures lacking comprehensive assessments of physical and transition risks may provide limited insights into potential impacts and vulnerabilities. Companies neglecting scenario analysis or stress testing may appear

Sector-specific examples further illustrate how best practices can be tailored to different industries, highlighting the importance of clarity, alignment with frameworks, and integration into business strategy. By adopting these best practices, companies can enhance their climate risk disclosures and better address stakeholder expectations and regulatory requirements.

7.9 Challenges and Barriers to Effective Climate Risk Disclosure

Despite its significance, effective climate risk disclosure faces challenges:

- **Data Availability and Quality**: Companies struggle with obtaining comprehensive climate risk data, affecting the accuracy of assessments (Griffin and Jaffe 2021).
- **Standardization Issues**: Multiple frameworks can create confusion and complicate reporting.
- **Balancing Transparency and Competitiveness**: Companies must manage proprietary information concerns while ensuring sufficient disclosure (Arian and Sands 2023).

7.10 Future Directions and Emerging Regulatory Trends

Anticipated Changes: Regulations are expected to expand, enhancing standards, integrating climate disclosures with financial reporting, and aligning with global frameworks like TCFD (Brown and Wald 2023).

7.10.1 Implications for Corporations

- **Increased Reporting Burden**: Companies may face greater demands for data collection and reporting.
- **Enhanced Focus on Climate Strategy**: Organizations will need to refine their climate strategies and set ambitious targets.
- **Opportunity for Leadership**: Companies excelling in climate risk disclosure can gain a competitive advantage.

7.10.2 Technological Innovations in Climate Risk Disclosure

Big data and artificial intelligence are transforming climate risk disclosure, enhancing the capabilities of data analytics, and predictive modeling. Such technologies enable better risk characterization, more informed decision making for strategic action, and hence greater quality of climate risk disclosure (Kumar and Khanna 2023).

- **Blockchain for Traceability**

  Blockchain enhances the intrinsic reliability and traceability of climate risk disclosure by providing a distributed, unalterable record. Thanks to a protected platform, climate relevant data can be made and recorded on site, i.e., output data and compliance reports additionally time-stamped and safely saved. Resistance to outside control ensures the precision and consistency of reported results, thus providing the confidence that is desirable to stakeholders. In addition, blockchain offers proof and audit trail of data sources and data transformations across time, which can be used to enable higher degrees of accountability.
- **Digital Reporting Platforms**

  With the development of digital reporting platforms, the management and delivery of climate risk information has changed completely. These platforms make the whole reporting process easier by providing their clients with means for composing reports, gathering data automatically and meeting the client's disclosure requirements. By streamlining these functions, they eliminate the difficulties and administrative burden that is typically linked with climate risk reporting. Such components as dashboards, graphics and automated computation allow for instant revision making disclosure even more appealing to the stakeholders. Companies also have the means to adhere both to the legal requirements and the best practices of the industry owing to the compliance checking and validation devices provided.

7.11 Role of Assurance and Verification

Third-party certification is a major instrument to advance the trustworthiness of climate risk disclosure. An external audit, which is no subjective evaluation of the accuracy and completeness of disclosure of a company on climate-related issues, guarantees compliance with the legal standards. This workflow reveals deficiencies that are otherwise invisible by internal review and, as such, preserves confidence of investors, regulators, and others. Assurance standards define the minimum that climate disclosures will be tested against to bring standardization and reliability. The key standards include:

- **ISAE 3000**: Guidelines for assuring non-financial information.
- **GRI Standards**: Focused on sustainability reporting, requiring accurate and complete disclosures.
- **SASB Standards**: Emphasizing industry-specific disclosures.

- **ISO 14064**: Guidelines for quantifying and reporting greenhouse gas emissions.

These assurance engagements involve thorough reviews of data collection processes, internal controls, and reporting practices, ensuring transparency and accountability.

Assurance processes are fundamental to generating confidence among a broad range of stakeholders (i.e., investors, regulators, customers and society). Companies that are adopting third-party confirmation and meeting assurance criteria confirms a stance on transparency and accountability that underpins the work in climate risk management. As a result, this independent corroboration of climate claims therefore improves investor relations, promotes sustainable and green capital, and improves corporate reputation.

7.12 Conclusion

The climate risk disclosure landscape is changing, with stakeholder pressures for global standards being reflected in all parts of the world, such as the International Financial Reporting Standards Foundation (IFRS) ISSB project which hopes to achieve global consistency in sustainability reporting standards (Bell and Handford 2023). Reported possible changes in include reframing of measures and methods in line with so-called what-if and scenario-based data i.e., in order to provide more information and insight on how stakeholders will be prepared to respond to climate risk in future. Sector-specific guidance will also become increasingly critical (Zhou 2022).

As climate risk gets integrated into broader reporting frameworks, environmental, social, and governance risk reporting has also found its practical implication. Businesses are urged to combine climate risk disclosures with all ESG disclosures to give stakeholders a holistic view of the credibility of sustainability performance claims made. New policies such as SFDR and EU Taxonomy push the boundaries of the integration as it demands high levels of data collection sophistication. It is important to mention frameworks such as the Task Force on Climate-related Financial Disclosures (TCFD), Global Reporting Initiative (GRI), and Sustainability Accounting Standards Board (SASB)" which have clear recommendations that enhance uniformity and comprehensiveness in climate reports. A sector-based approach understands that climate related risks are different for every sector and so is the required disclosure practices, thus improving the relevant sectoral insights for the stakeholders. New technologies like big data analytics, artificial intelligence (AI), and blockchain are improving the quality of climate risk assessment, credibility of data and efficiency in reporting, hence, companies are able to provide timely and accurate disclosures (Liu et al. 2024). Still, challenges and barriers exist, particularly with regards to the availability, and standardization of data, as well as the implicit trade-offs between disclosure and competitive advantage (Zhou 2022). These barriers have to be dealt with if the quality of disclosure is to be improved. The concern of investors appears to be affecting climate risk disclosure practices in so far as there is the erosion of trust and heightened concerns about disclosure requirements.

Institutional investors, for example, are also grumbling that there is a lack of more standard disclosure which would enable climate-related risk and opportunity to be assessed. In addition, practice disclosure is also motivated by external stakeholders (e.g., customers, local communities) who want companies to adjust their disclosure practices.

In summary, an effective climate risk disclosure is treated as an essential component for the governance and strategy. Adoption of standardized frameworks embraced with technology, and proper engagement of the stakeholders, can improvise the quality and efficacy of the climate related disclosures, which will not only benefit individual companies but also contribute to overall sustainability developments and advancements. This shift marks significant progress toward a more resilient, sustainable corporate landscape.

Conflict of Interest There is no conflict of interest for this research.

References

Aleksei C (2024) Regulatory frameworks for sustainable entrepreneurship in the arctic: environmental law and policy in the Russian Federation. BIO Web Conf

Arian A, Sands JS (2023) Corporate climate risk disclosure: assessing materiality and stakeholder expectations for sustainable value creation. Sustain Acc Manage Policy J

Bell M, Handford M (2023) The CFO imperative series-how will understanding climate risk move you from ambition to action? EY Climate Risk Disclosure Barometer

Ben-Ama W, Herrera DC, Martinez I (2023) Do climate risk disclosures matter to financial analysts? J Bus Fin Acc

Brown JR, Wald E (2023) Chilling climate change disclosure: the enabling role of corporate counsel in management misstatements of ESG matters. DePaul Law Rev. U Denver Legal Studies Research Paper

Day T et al (2022) Corporate climate responsibility monitor. New Climate Institute, Germany

Ennew C, Waite N, Waite R (2024) Financial services marketing: a guide to principles and practice, 4th ed. Routledge

Flammer C (2023) ESG and sustainable investing. National Bureau of Economic Research (NBER), Cambridge

Griffin P, Jaffe AM (2022) Challenges for a climate risk disclosure mandate. Nat Energy 7:2–4 (2022). https://doi.org/10.1038/s41560-021-00929-z

Hope A, Laasch O (2024) Responsible business: foundations of ethical and sustainable management, 3rd ed. Routledge

Ilhan E, Krueger P, Sautner Z, Starks LT (2023) Climate risk disclosure and institutional investors. Rev Fin Stud

Kumar R, Khanna R (2023) RPA (robotic process automation) in finance & accounting and future scope. In: 2023 Second international conference on augmented intelligence and sustainable systems (ICAISS). IEEE, pp 1640–1645

Kumar R, Kandpal B, Ahmad V (2023) Industrial IoT (IIOT): security threats and countermeasures. In: 2023 International conference on innovative data communication technologies and application (ICIDCA). IEEE, pp 829–833

Kumar R, Lande A, Kumar D, Malhotra RK, Sharma A (2023b) Technology bridging-in entrepreneurs and consumers in product development. In: 2023 11th International conference on intelligent systems and embedded design (ISED). IEEE, pp 1–5

Kumar R, Malholtra RK, Grover CN (2023c) Review on artificial intelligence role in implementation of Goods and Services Tax (GST) and future scope. In: 2023 International conference on artificial intelligence and smart communication (AISC). IEEE, pp 348–351

Kumar R, Saxena A, Singh R (2023d) Robotic process automation bridge-in banking institute and consumers. In: 2023 International conference on disruptive technologies (ICDT). IEEE, pp 428–431

Kumar R, Sexena A, Gehlot A (2023e) Artificial intelligence in smart education and futuristic challenges. In: 2023 International conference on disruptive technologies (ICDT). IEEE, pp 432–435

Kumar R, Goel R, Singh T, Mohanty SM, Gupta D, Alkhayyat A, Khanna R (2024a) Sustainable finance factors in Indian Economy: analysis on policy of climate change and energy sector. Fluct Noise Lett 23(2)

Kumar V, Kumar R, Goel R, Singh T, Kaur MJ (2024b) The evolution of digital marketing technologies: a fuzzy TOPSIS approach to selected technology in E-commerce platform. Int J Syst Assurance Eng Manage:1–14

Lei L, Zhang D (2024) ESG and corporate climate actions. In: Climate finance. Climate change and energy transition. Springer, Singapore

Liu Y, Xu Z, Xu Z, Zhu Y (2024) Can Chinese investors manage climate risk domestically and globally? Int Rev Econ Fin

Marković, V., Nagy, I., Sik, A. et al. Assessing drought and drought-related wildfire risk in Kanjiza, Serbia: the SEERISK methodology. Nat Hazards 80:709–726 (2016). https://doi.org/10.1007/s11069-015-1991-4

Mohanty SN, Singh T, Goel R, Baral SK, Kumar R (2024) A study on building awareness in cyber security for educational system in India using interpretive structural modellings. Int J Syst Assurance Eng Manage:1–11

Ngo T, Le T, Ullah S et al (2022) Climate risk disclosures and global sustainability initiatives: a conceptual analysis and agenda for future research. Bus Strat Environ

Norton GE, Leib EM, Plekenpol R (2024) Reducing food waste: emissions, environmental, and food waste reporting policies. The Global Food Banking Network, Cambridge

PwC (2022) PwC network-climate-related disclosures report. https://www.pwc.com/gx/en/corporate-responsibility/

Venturelli A, Mio C (eds) (2024) The Routledge handbook of accounting for the sustainable development goals, 1st ed. Routledge

Vestrelli R, Colladon AF, Pisello AL (2024) When attention to climate change matters: The impact of climate risk disclosure on firm market value. Energy Policy

Wu F, Zhang D, Ji Q (2024) Climate finance. Springer

Yoo M (2023) The spark of sustainability: analyzing the energy market's response to climate disclosure frameworks. Department of Environmental Studies

Zhang J, Song Z, Nedopil C (2024) China green finance status and trends 2023–2024. Griffith Asia Institute, Griffith University (Brisbane) and Green Finance & Development Center, FISF Fudan University (Shanghai). https://doi.org/10.25904/1912/5205

Zhou S (2022) Reporting and assurance of climate-related and other sustainability information: a review of research and practice. Aust Acc Rev

Chapter 8
Sustainable Finance and Sustainable Development Goal (SDG-2030)

Vasim Ahmad, Amar Johri, Rakesh Kumar, Firoz Alam, Umme Hani, and Anu Sayal

Abstract The chapter highlighted the pivotal role that sustainable finance plays in the realization of the United Nations' Sustainable Development Goals. The SDGs, which include 17 very interlinked global goals, are a universal framework to help combat the most unprecedented challenges facing the world, such as poverty, inequality, climate change, and environmental degradation, by 2030. In its entirety, sustainable finance incorporates ESG criteria in financial decision-making to mobilize capital for projects with positive social and environmental outcomes. This is the chapter about how sustainable finance converges with the SDGs, using financial instruments and strategies containing Green Bonds, Social Bonds, and Sustainability Linked Loans, among others. It shall go into the nuances of a proper understanding of the Sustainable Development Goals, historical background on the adoption of these goals, and their interlinkages among the goals. It will also focus on the mechanisms by which sustainable finance can deliver on the SDGs, covering the contribution of different players—such as governments, the private sector, and NGOs. Where

V. Ahmad (✉)
Uttaranchal Institute of Management-Uttaranchal University, Dehradun, India
e-mail: drahmadvasim@gmail.com

A. Johri · F. Alam · U. Hani
College of Administrative and Financial Sciences, Saudi Electronic University, Riyadh, Saudi Arabia
e-mail: a.johri@seu.edu.sa

F. Alam
e-mail: f.alam@seu.edu.sa

U. Hani
e-mail: uhani@seu.edu.sa

R. Kumar
Graphic Era Deemed to Be University, Dehradun, India
e-mail: aigrakeshkumar@gmail.com

A. Sayal
School of Accounting and Finance, Taylor's Business School, Taylor's University, Subang Jaya, Malaysia
e-mail: anu.sayal@taylors.edu.my

R. Kumar (ed.), *Sustainable Finance*, Sustainability Solutions,
https://doi.org/10.1007/978-3-031-97962-0_8

possible, it will also present successful project and initiative cases that have been effectively financed to lend support to the SDGs, complete with impact assessments and lessons learned. The indicators and metrics in this chapter address sustainable finance, measure the standards and frameworks in reporting, and the challenges in measurement and transparency. It identifies the barriers to financing SDG-related projects, opportunities for innovation and growth, and provides recommendations for policies that create enabling environments.

Keywords Sustainable development goals (SDGs) · Sustainable finance · Environmental · Social · And governance (ESG) criteria · Green bonds · Social bonds · Impact investing · Public–private partnerships (PPPs) · Financial instruments

8.1 Introduction

The input of 17 interlinked global goals is the Sustainable Development Goals, aiming to be a "blueprint to achieve a better and more sustainable future for all." By 2015, the United Nations had adopted the SDGs as part of the 2030 Agenda for Sustainable Development, purposed to address very serious current global challenges involving poverty, inequality, climate change, environmental degradation, peace, and justice. Most of the goals are accompanied by specific targets and indicators to help in guiding progress in implementation, thus ensuring holistic achievement of sustainable development (Solar Sister 2018). Sustainable finance is key to attaining SDGs. That means the incorporation of ESG into the processes of financial decision-making so that differentials between social and environmental impacts, already induced by several projects or activities, guide capital toward the right outcome in a differentiated manner. It mobilizes financial investments from both private and public resources to fill the funding gap created by sustainable development across boundaries, projected to be on the order of trillions of dollars every year (Bolton et al. 2020). Instruments and strategies within sustainable finance include green bonds, social bonds, sustainability-linked loans, ESG investing, and impact investing. These are financial instruments that mobilize money for such projects that fall under the provisions of the SDGs: renewable energy, sustainable agriculture, clean water and sanitation, affordable housing, and healthcare. Through transparency and accountability, value creation over the long time, and control of the risks associated with climate change, social inequalities, and failures in governance, it helps to ensure a more resilient and inclusive global economy (OECD 2017).

Figure 8.1 provides the 17 Sustainable Development Goals (SDGs) established by the United Nations.

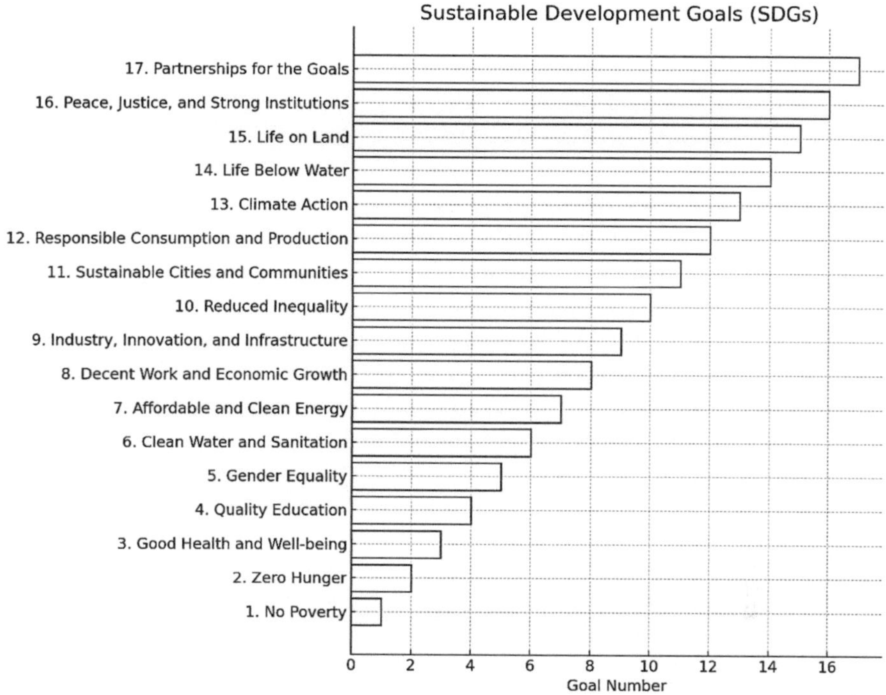

Fig. 8.1 17 Sustainable development goals (SDGs) established by the United Nations

8.1.1 Historical Background and Adoption by the United Nations

The United Nations General Assembly adopted the Sustainable Development Goals in September 2015 as part of the 2030 Agenda for Sustainable Development. It is a universal framework that is all-inclusive to sustainably steer the efforts of the globe to year 2030. The SDGs succeeded the MDGs, that were set up in 2000 to address problems such as poverty, hunger, and disease (ICMA 2020).

Though developmentally successful in many ways, the MDGs did have some tags on their potential to move forward—geographic, toward developing countries; sectoral, toward specific sectors. Feeling this gap, the SDGs were drafted to be inclusive and holistic, putting a strong line on the correlatedness of economic, social, and environmental sustainability. This shift recognizes that sustainable development must tackle the multidimensional human and environmental well-being through an integrated approach (Ceres 2018). These Sustainable Development Goals were preceded by broad-ranging consultations and negotiations among member states, NGOs, and other stakeholders to that effect. In these processes, there was an identified need for more inclusive and participatory approaches toward global development, so that no one would be left behind. The SDGs were formally adopted at the United Nations

Summit on Sustainable Development held in New York between September 25 and 27, 2015, and entered into force as of January 1, 2016. (UNEP Finance Initiative 2018).

8.1.2 Interlinkages Among the SDGs

The SGDs view these goals as interlinked, recognizing that progress in one area often depends on strides in others. For instance, in regard to gender equality (Goal 5), it can be a good reason for the achievement of learning outcomes for education (Goal 4), health (Goal 3), and growth of the economy (Goal 8). It is with this sense that climate action (Goal 13) is central to sustainable cities and communities (Goal 11), life below water (Goal 14), and life on land (Goal 15).

Interlinkages of the SDGs underline the need for synergized approaches to policy-making and implementation. For instance, sustainable agriculture ensures clean water and sanitation through better water resources management and decreases climate change impact. In the same way, sustainable economic growth and innovation help build resilient infrastructure and lead to reduced inequalities. Policymakers find an understanding of these interconnections decisive in order to be able to develop strategies maximizing synergies and minimizing trade-offs. The very reconciliatoriness of the SDGs is that efforts made in one of the goals spur the achievement of others, triggering an even more sustainable and inclusive overall global development path (Global Reporting Initiative [GRI] 2020).

Figure 8.2 illustrate some of the key interlinkages among them, demonstrating the integrated nature of the goals and the necessity of coordinated efforts to achieve sustainable development.

8.1.3 Objectives of the Chapter

The primary purposes of the chapter are to discuss and explain the multi-relationship between sustainable finance and the SDGs. Concretely, the chapter intends to:

1. Deepen understanding on the SDGs: Devote some time to explaining each of the 17 SDGs in detail, including a little history, adoption at the United Nations, and how they are all interlinked.
2. How sustainable finance aligns with the SDGs: How, through very numerous mechanisms and strategies, has it contributed to the achievement of the SDGs? This shall involve talking about the roles played by various stakeholders: policy actors like governments, individual companies in the private sector, and NGOs.
3. Financial instruments and tools: Financial instruments, such as green bonds, social bonds, and sustainability bonds, would be explored alongside ESG and

Primary SDG	Linked SDGs	Explanation
Goal 2: Zero Hunger	Goal 6: Clean Water and Sanitation	Sustainable agriculture improves water management
	Goal 13: Climate Action	Sustainable farming practices reduce impact on climate change
Goal 5: Gender Equality	Goal 3: Good Health and Well-being	Gender equality leads to better health outcomes for women and children
	Goal 4: Quality Education	Empowered women are more likely to ensure education for their children
	Goal 8: Decent Work and Economic Growth	Gender equality in the workforce boosts economic productivity
Goal 7: Affordable and Clean Energy	Goal 13: Climate Action	Renewable energy reduces greenhouse gas emissions
	Goal 9: Industry, Innovation, and Infrastructure	Access to energy supports industrial innovation and infrastructure development
Goal 11: Sustainable Cities and Communities	Goal 6: Clean Water and Sanitation	Sustainable urban planning includes effective water management systems
	Goal 13: Climate Action	Resilient cities can better cope with climate-related disasters
Goal 15: Life on Land	Goal 13: Climate Action	Forest conservation and reforestation help mitigate climate change
	Goal 6: Clean Water and Sanitation	Healthy ecosystems ensure clean water supply
Goal 8: Decent Work and Economic Growth	Goal 9: Industry, Innovation, and Infrastructure	Economic growth is driven by industrial innovation and robust infrastructure
	Goal 10: Reduced Inequality	Inclusive economic growth reduces inequalities within and among countries
Goal 16: Peace, Justice, and Strong Institutions	Goal 10: Reduced Inequality	Just and inclusive societies are essential for reducing inequalities
	Goal 17: Partnerships for the Goals	Effective institutions are critical for fostering global partnerships and implementing sustainable development

Fig. 8.2 Interlinkages among selected SDGs

impact investing; at the same time, PPPs in terms of their role in financing sustainable development projects would be covered.

4. Case Studies Presentation: Sharing real experiences about the success stories of projects and initiatives that have been financed toward supporting the SDGs, assessing the impact of such projects, and lessons learned.
5. Measurement and Reporting: Relate the metrics and indicators used as measurements for sustainable finance, frames, and standards of reporting, and challenges in measurement and transparency.
6. Challenges and Opportunities: The constraints of financing based on issues pertinent to SDGs, opportunities for innovations and growth, and policy recommendations that make enabling environments in sustainable finance.
7. Provide future perspectives: Use your analysis above to predict new trends and innovations of sustainable finance covering the SDGs, and try to imagine what the future of this field might look like.

8.2 Alignment of Sustainable Finance with SDGs

Second, sustainable finance plays a prominent role in driving progress toward the achievement of the SDGs. It does this by channeling investments into projects and initiatives that foster economic, social, and environmental sustainability. It serves the SDGs in a number of ways:

1. Resource Mobilization: Sustainable finance mobilizes and channels private capital for social good into sectors like renewable energy, sustainable agriculture, and social infrastructure all directly linked to achieving the SDGs (European Commission 2020).
2. Risk Mitigation: Using the criteria of ESG, sustainable finance can mitigate risks related to climate change, social inequalities, and bad governance; such a long-term view creates stability and provides long-term returns for investments (Banga 2019).
3. Innovation and Inclusion: Investment in green technology and business models can act as a means to an end of innovation and inclusivity, opening up opportunities for the marginalized communities to grow inclusively (United Nations 2018).

8.2.1 Mechanisms and Strategies to Align Investments with SDGs

A mix of regulatory frameworks, market-based instruments, and innovative financing mechanisms is required in order to align investments with the SDGs.

1. Regulatory Frameworks: At the very core, very critical are the roles that governments play in setting regulations and policies that foster sustainable investing. These differ from carbon pricing to tax incentives for green investments, progressing on to mandatory disclosure requirements on ESG performance, among many other factors as indicated by WBCSD in 2018.
2. Green Bonds and Social Bonds: These are specialized financial instruments used to fund projects with positive impact on the environment or society (Sachs 2012). For instance, Green bonds would finance renewable energy projects; at the same time, Social Bonds would provide funds for low-cost housing or projects that deal with diseases (World Economic Forum 2018).
3. Sustainability-Linked Loans: These loans offer better terms depending on the performance of the borrower against predetermined sustainability targets, hence incentivizing companies toward better ESG practices (Fink 2020).
4. Impact Investing: The investors actually look to produce measurable social and environmental benefits alongside financial returns. This investment is mostly done in social Enterprises, Community Development Finance Institutions, etc., while mapping projects against SDGs (Jack and Suri 2011).

8.2.2 Role of Stakeholders

Sustainable finance has its paths for the SDGs. It engages many stakeholders playing distinct, albeit complementary, roles:

1. Governments: They hold the reins for setting regulatory environments that support sustainable investments in their totality. They can offer incentives to green investments, establish public–private partnerships, and assure alignment of public expenditure with the SDGs as a whole (CDP 2019).
2. Private sector: Corporations and financial institutions play significant roles in incorporating ESG criteria into the investment decisions and manner of conducting business. Adoption of models of sustainable businesses can be their gigantic contribution to meeting the SDGs (Global Impact Investing Network [GIIN] 2020).
3. NGOs: Their advocacy role is with regard to the practices of sustainability, holding stakeholders to account, and providing first-hand, on-the-ground views into community needs. Rather often, they facilitate public–private partnerships and are, in part, a way to track progress toward the attainment of the SDGs (Le Blanc 2015).

8.3 Financial Instruments and Tools

8.3.1 Green Bonds

These are debt instruments that aim at funding projects that have a positive impact on the environment through the provision of an adequate source of financing. Such projects include, among others, renewable energy, enhancement of energy efficiency, waste management that is sustainable, and conservation efforts. Green bonds have had significant growth in both their investors and issuers identifying the daunting issues of mitigating climate change and promoting sustainability (Task Force on Climate-related Financial Disclosure 2017).

Some of the benefits that green bonds offer includes the following: Environmental Impact: Green bonds finance or refinance projects that do not affect the environment negatively, thus reducing global warming and enhancing sustainable development. Investor Appeal: It appeals to a new class of investors whose demand is fast on the rise and looking to invest in sustainable initiatives with win–win possibilities of reaping financial results. Market Growth: From Busch and Lewandowski 2018, the market for green bonds is on the rise, and for issuers, this opens more avenues to raise financing for their environmental projects (Busch and Lewandowski 2018).

8.3.2 Social Bonds

These are similar to green bonds, with the difference being that they finance projects which yield positive social outcomes. One such undercurrent relates to low-cost housing, health, education, and community development. In other words, they try to improve social welfare and achieve all-inclusive growth. The major advantages of social bonds are that they: Create Social Impact: They provide the much-needed funding for projects that resolve major social issues and benefit underprivileged people or society (Amel-Zadeh and Serafeim 2018). Diverse Investment Options: Social bonds give investors a chance for diversification in their portfolios, supporting socially beneficial activities. Alignment with the SDGs: Social bonds directly contribute to the achievement of a number of the Sustainable Development Goals, such as quality education, good health and well-being, and reduced inequalities, among others [UNDP] (2016).

8.3.3 Sustainability Bonds

These are such bonds that combine green and social bond elements to finance projects with environmental and social benefits. The bonds finance/support comprehensive initiatives that attain several dimensions of sustainability as the result of high interconnection of the SDGs. According to Moody's Investors Service, 2020, some of the benefits of sustainability bonds are:

Holistic approach: Sustainability bonds go further to provide a more integrated approach in addressing environmental and social issues and thus holding onto sustainable development.

Broad Investor Base: They attract investors interested in comprehensive sustainability strategies, enhancing market depth and liquidity (Krueger et al. 2020).

Increased Impact: Truly, sustainability bonds are all about raising finance for projects with positive impacts that diffuse into society across sectors and communities.

8.3.4 ESG Investing and Impact Investing

ESG Investing: Investment process unifying the concepts of environmental, social, and good corporate governance factors into a single holistic process seeking to create long-term financial returns and positive societal impact. Some of the most common ESG factors include:

Environmental: Climate change mitigation, resource efficiency, and reduction of pollution. Social: Labor practice, community engagement, and human rights.

Governance: Board diversity, corporate ethics, transparency. UNEP, 2019. ESG investing has been gaining momentum because the investors see a connection between sustainable business practice and a company's financial performance. There are a number of key strategies in ESG investing:

Screening: Excluding companies that do not meet certain ESG criteria—negative screening—or including those that excel in ESG performance (positive screening) (Task Force on Climate-related Financial Disclosures 2018).

Integration: Integrated ESG to assess the risk and opportunity versus traditional financial analysis.

Engagement: Actively engaging with companies on ESG practices and performance improvement (United Nations 2015a, b).

8.3.5 Impact Investing

Impact investing is the combination of generating measurable social and environmental impact along with financial returns. Where ESG investing looks to sustainable practice for risk mitigation and performance enhancement, impact investing is about directly achieving certain outcomes pertaining to the SDGs (Morgan 2020).

Some defining features of impact investing include intentionality, measurability, and additionality. The investment must be made with the expressed intention of creating a positive impact. It should also measure and report the outcomes or impact created by the investment on society and environment. Additionally, impact investments should achieve additionality, that is, they should result in positive impact that would not have been created anyway without such investment. Impact investments cut across asset classes and sectors such as clean energy, healthcare, education, low-cost housing investing directly in social enterprises and community development finance institutions (Asif et al. 2013).

8.3.6 Public–Private Partnerships (PPPs)

Public–Private Partnerships are cooperative arrangements between public authorities and private firms to provide financing, construction, and then operation of projects in the public interest. The very objective of PPP is to mobilize resources in such a way that it fosters sustainable development, capitalizing on the strengths of both parties and pooling their resources (Robeco 2019). Resource Mobilization: PPPs mobilize private capital and expertise for public projects, reducing the financial burden on governments. Efficiency and Innovation: Involvement of the private sector lends efficiency, innovation, and managerial expertise to the implementation and operation of projects. Risk Sharing: PPPs provide for risk-sharing between public and private partners involved in a project w, therefore enhancing its feasibility and sustainability. In this regard, PPPs may make valuable contributions toward the realization of the Sustainable Development Goals in the following way: Infrastructure development via the financing and development of crucial infrastructure for transportation, energy, and water systems is key in achieving sustainable cities and communities, or SDG 11.

8.4 Case Studies

8.4.1 Grameen Bank Microfinance Project (SDG 1: No Poverty)

Grameen Bank—founded by Nobel laureate Muhammad Yunus, this organization provides microloans to poor people, mostly women, in Bangladesh. The basic idea is to help them overcome poverty by starting small businesses that would make the owners financially self-reliant. More than 9 million members get benefited, leading in the translation of millions of people out of poverty. Ninety-seven percent of the borrowers in this bank are women, and its social impacts are very great. Economic Development: The project has boosted local economies by nurturing entrepreneurship and job creation (Yunus 2007).

8.4.2 The Bujagali Hydropower Project (SDG 7: Affordable)

The Bujagali Hydropower Project in Uganda is a 250 MW hydropower plant. The project is aimed at providing the country with sustainable, affordable energy (Lamberton 2005). In 2005, the financing of this project had been fully funded by the public and private sectors because of finance investments from IFC and other DFIs.

Energy Access: Through the provision of electricity to over 3 million Ugandans, energy poverty is being diminished. Economic Growth: Reliability in the supply of electricity propels growth industrially and betters the offer of public services. Environmental Sustainability: The project built its policy on renewable energy; this policy is cutting greenhouse gases and has reduced their dependence on fossil fuels (World Economic Forum 2020).

8.4.3 The M-Pesa Mobile Money Service (SDG 9: Industry, Innovation, and Infrastructure)

M-Pesa is the facility of mobile payment transfer inaugurated by Safaricom in Kenya, used by people to deposit, withdraw, transfer money, and pay for services and goods with their mobile phones. It has made any sort of financial involvement in Kenya, and above all, most of the least-developed countries, very different. Financial Inclusion: Now, this has 30 million plus subscribers, mostly those who never had bank accounts, to avail financial services. Economic Empowerment: Prompted small business transactions and personal remittances, enabling economic empowerment. Innovation: Helped breeding more of the same, in mobile banking services globally, hence contributing to financial innovation (S&P Global 2020).

8.4.4 Amazon Rainforest Conservation Project—SDG 15: Life on Land

The program works in the domain of sustainable land management, reforestation, and support to indigenous communities for Amazon rainforest conservation. It is financed using a blend of international donors, NGOs, and government programs (Erlandsson and Barchiche 2019). Biodiversity Conservation: Conserved protected millions of hectares of rainforest, saving countless species. Community Development: Provided indigenous communities with the power of sustainable livelihoods and advocacy of land rights. Climate Mitigation: Helped reduce deforestation rates as a contribution to fighting global climate change. (World Bank 2016a, b).

Success Stories and Impact Assessments

1. ***Success of the Solar Sister Project (SDG 5: Gender Equality and SDG 7: Affordable and Clean Energy)***

It empowers women in sub-Saharan Africa by training them as entrepreneurs in solar energy. Through the use of training and solar products, women can sell these products within respective communities. Therefore, Solar Sister not only improves access to clean energy for low-income off-grid communities but also furthers the cause of women empowerment. This is an important program in many ways because current statistics show that women in sub-Saharan Africa bear a disproportionate part of the world's energy burden. Accordingly, mission accomplishment on the company's part infuses the dual benefit of making clean energy available to people at the base of the economic pyramid while empowering women. Health and Environment: Reduced indoor air pollution and associated carbon emissions lead to better health and more sustainable environments (Solar Sister 2018). Economic Impact: The enhanced income for women entrepreneurs has had spillover benefits to improve local economies. Social Impact: Improved social status and enhanced decision-making powers in households and communities for the concerned women (UNEP Finance Initiative 2019).

2. ***The Impact of the Kigali Cooling Efficiency Program (SDG 13: Climate Action)***

The Kigali Cooling Efficiency Program develops country markets, improves the energy efficiency of cooling systems, and reduces their GHG emissions. It provides funding and technical assistance to facilitate the uptake of energy-efficient cooling technologies in intervention countries. Climate Mitigation: Huge GHG savings from improved energy efficiency (Eccles et al. 2014). Economic Savings: Reduced energy costs for consumers and businesses by enhancing resilience to the economic impacts. Health Advantages: More excellent indoor air quality; fewer heat-related health issues. Environmental Effects: Likely to prevent billions of metric tons of CO_2-equivalent from being emitted by 2050. Economic Effects: Could save billions of dollars on energy costs, reducing a load on the wallet of developing economies (Centre for International Environmental Law [CIEL] 2018).

8.5 Measurement and Reporting

Such will require a strong set of metrics and indicators to account for the impact of initiatives in sustainable finance and their relation to attaining SDGs. Metrics and indicators help quantify progress and ensure accountability. Key metrics include.

8.5.1 Environmental Metrics

Greenhouse Gases: Metric tons of CO_2 equivalent of reduced or avoided GHG emissions.

Energy Use: The quantity of renewable energy produced or consumed.

Water Usage: Volume of water saved or sustainably managed.

Biodiversity Impact: Area of natural habitats protected/restored, e.g., hectares of forest conserved.

8.5.2 Social Metrics

Poverty Reduction: The number of people moved out of poverty or access to financial services.

Health and Well-being: Improving health outcomes, such as reduced mortality rates and increased access to health care (Kölbel et al. 2020).

Education: Proportion of population accessing quality education or training.

Gender Equality: Indicators of gender equality in employment, income, and education.

8.5.3 Governance Metrics

Board Diversity: proportion of women and other underrepresented groups who sit on the corporate board.

Transparency and Accountability: Reliability and competence of ESG disclosures.

Compliance and Ethics: Number of compliance violations or ethical breaches.

8.5.4 Reporting Standards and Frameworks

There have been several standards and frameworks developed to help organizations with measuring and reporting their sustainability impacts, among which the most prominent includes:

1. Global Reporting Initiative: The GRI Standards are the most globally applied standards for sustainability reporting. They provide the reporting framework that organizations can use to report on their economic, environmental, and social impacts. GRI Standards underline transparency and accountability. That is to say, customers are in a place to scrupulously evaluate the organizational performance based on certain formulated criteria (Refinitiv 2020).
2. Sustainability Accounting Standards Board (SASB): SASB offers industry-specific standards for other entities voluntary to disclose reportable sustainability materials likely to impact on the results or outcomes of financial activity (PwC 2019). The materiality of the standards is created to be of great use to the investors always, as it focuses on information that can be material financially. SASB has standards for reporting in different sectors; they provide ready guidance on how to report on the sustainability performance.
3. Task Force on Climate-related Financial Disclosures (TCFD): The TCFD framework is viewed as one providing guidance on climate-related financial disclosures. The TCFD framework revolves around making voluntary reporting recommendations on climate-related risks and opportunities. The major objective of the framework from the TCFD is to increase transparency on the impacts of climate on an organization's operations, consequently helping investors better price and assess the said risks and opportunities (World Resources Institute [WRI] 2019).

8.5.5 Challenges of Measurement and Transparency

Despite the availability of metrics and frameworks, several challenges persist in measuring and reporting sustainability impacts:

1. Data availability and quality: Inconsistent data: the data collection methods and quality applied by organizations are highly varied and differ from one region to another, hence the inconsistent nature of their reporting. Data gaps: in some areas, and especially in the developing countries, quite some data may not be readily available; hence, comprehensive impact assessment may be hampered (Lazard 2020).
2. Standardization Problems: Non-Uniformity: The variation in policies and frameworks for reporting creates confusion and often makes it hard to assess the business performance of an organization compared to others. Expanding Standards: Sustainability standards and best practices are ever expanding; thus, the expectation of organizations is an adjustment to any improvements, amendments, or accretion in reporting practice (Global Impact Investing Network [GIIN] 2020).
3. Complexity in Impact Measurement Multidimensional impact: It is innately complicated measuring multidimensional impacts on a sustainability initiative, either of an economic, social, or environmental nature and requires a very sophisticated methodology. Attribution and Causality—The identification of the direct contribution of certain initiatives and attribution of results to particular actions are quite complicated, in particular within systemic problems (Leventis et al. 2013).
4. Transparency and Accountability: Greenwashing: There is an organization's potential danger in the practice termed "greenwashing"—the process of giving a misleading record regarding their performance without substance. Trust by Stakeholders: Trust in sustainability reports rides based on robust (World Bank 2016a, b).

8.6 Challenges and Opportunities

1. Inadequate Access to Capital: Many projects that are series of the SDGs lack the needed capital. This is particularly the case for developing counties, where financial markets may be underdeveloped and investors often hold back on committing funds because of perceived risks. Improved access to affordable financing can diffuse or help overcome major implementation obstacles in projects contributing toward sustainable development (Principles for Responsible Investment [PRI] 2018).
2. Perceived Risks: Even another significant barrier to investments in SDG-linked projects by investors is the risk factor. This includes risks of political instability, regulatory uncertainty, and market volatility, underscoring that some may deter or discourage private investment and lead to over-reliance on public financing,

usually constrained in supply. In addition, most of the SDG-related projects are long-term and require patient capital, which happens to be in short supply (Global Reporting Initiative 2018).

3. Poor Measurement and Reporting: Another challenge to investing in SDGs is that the measurement and reporting of investment impact are complex. Further, difficult attribution of that specific outcome to a particular investment—together with inconsistent data and lack of standardized metrics —make it very hard for investors to assess whether these projects are really impactful or perform well. This can potentially decrease investor confidence and will to commit capital toward attaining the SDGs (United Nations Global Compact 2018).
4. Regulatory and Policy Constraints: The policy and regulatory environments of most countries are still not enabling or conducive to incentivizing investments in sustainable development. Incoherent policies, lack of clear regulatory frameworks, and poor incentives for sustainable business practices offer difficulties to investors and project developers. Moreover, policy instability and frequent changes in regulations can increase uncertainty and scare away investment (Garcia et al. 2017).

8.6.1 Opportunities for Innovation and Growth

1. Green Financial Instruments Development; Designing and increasing green financial instruments, including green bonds, social bonds, and sustainability-linked loans, present new avenues for the financing of SDG-related projects. They are financial instruments that attract a subset of fast-growing investors who aim at generating positive environmental and social impacts alongside financial returns. At high scale in mobilizing capital for sustainable development, reaching substantial such innovations in these financial products are possible (High-Level Expert Group on Sustainable Finance [HLEG] 2018).
2. Technological Advancements: Given technological advancements, there is great potential to solve SDG-related challenges. Innovations in renewable energy and sustainable levels of agricultural and water management practices, along with digital finance innovations, have the capacity to raise operational efficiency and effectiveness for delivery-based projects on the SDGs. For example, blockchain can increase transparency and traceability in supply chains, while "fintech" solutions can provide access to financial services for underserved populations better (Loan Market Association [LMA] 2019).
3. Blended Finance Models: One can use blended finance models that blend public and private funding to mitigate these risks associated with attracting private investment in projects relevant to SDGs. Public funds could be mobilized to decrease the risk associated with certain investments and hence make them more attractive to private investors. This may include mechanisms such as guarantees, first-loss capital, and concessional loans that align the interests between the public and private stakeholders combined within the models (K-CEP 2019).

4. Impact Investing: Imperatively, impact investing is a new field that pursues to achieve measurable social or environmental benefits alongside financial returns. This very approach goes very well with the ideals of attaining SDGs and opens up prospects for investors in projects that bring positive societal impacts. Increased growth in impact investing could channel or garner massive resources toward the attainment of the SDGs and help accelerate innovation and sustainable development further.—United Nations Conference on Trade and Development [UNCTAD] (2014).

8.6.2 Policy Recommendations and Enabling Environments

1. Regulatory frameworks need to be stronger: Governments must build and enforce clear regulatory frameworks that help and support investments in sustainable development. It will include measures of obligatory requirements of reporting on sustainability, plus tax incentives for green investments while introducing carbon pricing. Consistent and predictable policies can support much-needed investor and project developer confidence. (Morgan 2020).
2. Strengthening Public–Private Partnerships: Public–private partnerships (PPPs) can play a very big role in the financing and the implementation of projects associated to SDGs. Governments, therefore, need to facilitate the formation of PPPs by providing cofinancing, mechanisms for sharing risks, friendly legal frameworks, and policies. The nurturing of the willingness of the private sector in collaborative efforts will bring more resources and expertise to bear on sustainable development (World Bank 2016a, b).
3. Financial Innovation: Ensure an enabling environment for innovation in finance and the utilization of tools and models for financing the needs of sustainable development in accordance with SDG requirements. Further, encourage more green bonds, social bonds, and sustainability-linked loans. Encourage blended finance and impact investment. Piloting and scaling up new financial product innovations can be simplified with regulatory sandboxes and innovation hubs (KPMG 2020).
4. Building Capacity and Raising Awareness: The capacity development is intended to provide the stakeholders with the necessary underpinning knowledge and skill so that they can conceive, implement, and finance projects that would be carried out within the SDG framework. This further includes training for government officials, financial institutions, and project developers on the principles and practices of sustainable finance. Moreover, awareness of the benefit that can be derived from sustainable investments should further do good by channeling more capital toward financing the SDGs (UNCTAD 2014).

8.7 Predictions and Future

The drive toward achieving the Sustainable Development Goals would be mainstreamed by the increasing criticality of sustainable finance. It is also expected that pension funds, insurance companies, and sovereign wealth funds, all being types of institutional investors, would start to allocate a higher portion of their portfolios in ESG-integrated investment instruments. It will be driven by the realization that sustainable practices can yield long-term financial stability and risk mitigation (BlackRock 2020). Impact investing is going to scale dramatically now as more investors seek to deliver measurable social and environmental impacts with financial returns: the market for impact investing will be in the hundreds of billions of dollars in the coming years. This will be backed up by the improvement in measures and reportage to show impact, which will make it easy for investors to access and measure outcomes from their investment (GIIN 2020). The future of sustainable finance lies in technological advancements including blockchains, artificial intelligence, and big data analytics that will be rather effective for expanding transparency, enhancing efficiency in the markets, and promoting accountability.

For example, using blockchain, immutable records of transactions could become available for traceability in sustainable investments, while artificial intelligence could be used to analyze large amounts of data to identify ESG-related risks and opportunities.

8.7.1 Emerging Trends and Innovations in Sustainable Finance Vis-à-Vis the SDGs

1. Green and Sustainable Bonds The green, social, and sustainability bond market is set to grow. These bonds are likely to find some special interest from the side of the investor, which provides some dual benefit in a financial return coupled with some good to the environment or society. Innovations also include sustainability-linked bonds whose coupon rates change with the accomplishment of marks on certain sustainability outlook by the issuer (ICMA 2020).
2. Blended Finance Models: This is a pathway where considerable interest is growing in deploying such a structure to de-risk investments in SDG-related projects. Blended finance sources public, philanthropic, and private capitals, combining such kinds of investment into one funding source. Such public and other philanthropic spending leverage private sector investment through first-loss or guarantee arrangements—a very effective way of raising money for hard-hitting projects in developing countries with a higher perceived level of risks.
3. ESG Integration and Standardization: There is an increasing trend that mainstream investment processes are now incorporating ESG factors. Asset managers

and investors are increasingly using ESG data to make informed financial investment decisions. Additionally, development focused on ESG integration standardization has attracted keen interest, including the Global Reporting Initiative, the Sustainability Accounting Standards Board, and the Task Force on Climate-Related Financial Disclosures, among others.
4. Emergence of Green Fintech: The confluence of finance and technology, or fintech, is creating a force for new innovation in sustainable finance. Green fintech in solutions that are suddenly emerging to deal with all sorts of sustainability challenges: for instance, developing digital platforms with a robust interface that would facilitate the entire process of green bond issuance and trading; on the other hand, a mobile application is hawking sustainable consumer behavior and mentoring by tracking personal carbon footprints and offering green investment solutions for consumers. (WEF 2020).

8.7.2 Policy Recommendations and Enabling Environments

1. Supportive Regulatory Frameworks: Governments need to create supportive regulatory frameworks so that they encourage sustainable investments. This would mean taking on policies that call for ESG disclosures, offer double tax benefits on green investments, and introduce carbon pricing mechanisms. Regular and predictable regulations may mitigate uncertainties and pull long-term capital toward potential sustainable projects (UNDP 2016).
2. Capacity Building and Education: Create a capacity base in financial institutions, regulators, and investors. The main principles and practices of sustainable finance education can be developed through training seminars that will likely raise awareness and knowledge, leading to market players being more effective. Governments and international organizations should invest in programs for building capacities for transition toward sustainable finance (UNCTAD 2014).
3. Public–Private Collaboration: Public–private collaboration, close cooperation, and a closer partnership among public, private entities, and members of the civil society are key and very important in scaling up sustainable financing. Public–Private Partnerships (PPPs) are mechanisms through which the comparative advantages and pooled resources from the private and public sectors come into play to finance and execute SDGs-related projects. Governments must thus play a role in supporting these PPPs with co-financing, risk-sharing mechanisms, and an enabling legal and institutional environment (World Bank 2016a, b).

8.8 Conclusion

Sustainable finance is essential in the advancement toward the achievement of the priorities set forth by SDGs. Sustainable finance mobilizes investments by financial entities toward projects that foster sustainable development of environmental, social, and governance dimensions. Some of the key financial instruments available in the market, such as green bonds, social bonds, and sustainability bonds, fund environmental and social projects coupled with ESG and impact investing with measurable positive impacts together with financial performance. This calls for the emphasis on the measurement and reporting for effective assessment of the impact of sustainable finance.

Old, established frameworks, like the Global Reporting Initiative and the Sustainability Accounting Standards Board, and emerging frameworks, like the Task Force on Climate-related Financial Disclosures, provide harmonized approaches to translate sustainability into drivers backed by integrated reporting. Many challenges remain: from data availability and reporting consistency to regulatory hurdles and enabling practice at all scales. The effort to work on measurable practices and enablement environments should be constantly improved. The Task Force takes on board that the outcomes of these efforts and the mainstreaming of sustainable finance in light of achieving the SDGs will leave no one indifferent. By this point, when urgency is constantly and persistently enveloping global challenges like climate change, poverty, and inequality, massive investment must be put into sustainable development. Sustainable finance will be where and how capital is mobilized and allocated toward projects and activities that are poised to make real epochal changes in society, to instill sustainable changes and improve the environment.

By integrating sustainability considerations into decision-making on the use of resources, investors, companies, and governments can help realize a more equitable and robust global economy. Sustainable finance does not only take care of immediate social and environmental issues but does also favor long-term economic stability and growth. It promotes the adoption of sustainable business practices, innovation, and building resilience both within human communities and ecosystems. Aligning the financial systems with the SDGs will be instrumental in building a path forward toward a sustainable future for sustainable finance. Indeed, binding efforts are required of governments, the private sector, non-governmental organizations, and the financial community. Working in a synchronized manner, we could virtually warrant the most effective deployment of financial resources to attain SDGs—making a world where economic prosperity, social inclusion, and environmental inclusivity are the implicit notions.

Funding NA.

Conflict of Interest: There is no conflict of interest for this research.

References

Amel-Zadeh A, Serafeim G (2018) Why and how investors use ESG information: evidence from a global survey. Financ Anal J 74(3):87–103. https://doi.org/10.2469/faj.v74.n3.2

Asif M, Searcy C, Zutshi A, Fisscher OAM (2013) An integrated management systems approach to corporate sustainability. J Clean Prod 56:7–17. https://doi.org/10.1016/j.jclepro.2011.10.034

Banga J (2019) The green bond market: a potential source of climate finance for developing countries. J Sustain Finance Investment 9(1):17–32. https://doi.org/10.1080/20430795.2018.1522167

Bolton P, Després M, Pereira da Silva LA, Samama F, Svartzman R (2020) The green swan: central banking and financial stability in the age of climate change. Bank for International Settlements. Retrieved from https://www.bis.org

Busch T, Lewandowski S (2018) Corporate carbon and financial performance: a meta-analysis. J Ind Ecol 22(4):745–759. https://doi.org/10.1111/jiec.12591

CDP (2019) Major risk or rosy opportunity: are companies ready for climate change? Retrieved from https://www.cdp.net

Centre for International Environmental Law (CIEL) (2018) Fuel to the fire: how geoengineering threatens to entrench fossil fuels and accelerate the climate crisis. Retrieved from https://www.ciel.org

Ceres (2018) Financing a sustainable future: How investors are measuring ESG performance. Retrieved from https://www.ceres.org

European Commission (2020) Renewed sustainable finance strategy and implementation of the action plan on financing sustainable growth. Retrieved from https://ec.europa.eu

Eccles RG, Ioannou I, Serafeim G (2014) The impact of corporate sustainability on organizational processes and performance. Manage Sci 60(11):2835–2857. https://doi.org/10.1287/mnsc.2014.1984

Erlandsson L, Barchiche D (2019) Sustainability-linked bonds: from theory to impact. World Resources Institute. Retrieved from https://www.wri.org

Fink L (2020) A fundamental reshaping of finance. BlackRock. Retrieved from https://www.blackrock.com

Garcia AS, Mendes-Da-Silva W, Orsato RJ (2017) Sensitive industries produce better ESG performance: evidence from emerging markets. J Clean Prod 150:135–147. https://doi.org/10.1016/j.jclepro.2017.02.180

Global Impact Investing Network (GIIN) (2020) Annual impact investor survey. Retrieved from https://www.thegiin.org

Global Reporting Initiative (GRI) (2018) GRI standards: reporting principles and standards. Retrieved from https://www.globalreporting.org

Global Reporting Initiative (GRI) (2020) GRI standards: reporting principles and standards. Retrieved from https://www.globalreporting.org

HLEG (2018) Financing a sustainable European economy: final report. High-level expert group on sustainable finance. Retrieved from https://ec.europa.eu

ICMA (2020) Green bond principles. International Capital Market Association. Retrieved from https://www.icmagroup.org

Jack W, Suri T (2011) Mobile money: the economics of M-PESA. National Bureau of Economic Research. Retrieved from https://www.nber.org

K-CEP (2019) Kigali cooling efficiency program annual report. Retrieved from https://www.k-cep.org

Kölbel JF, Heeb F, Paetzold F, Busch T (2020) Can sustainable investing save the world? Reviewing the mechanisms of investor impact. Organ Environ 33(4):554–574. https://doi.org/10.1177/1086026619857630

KPMG (2020) The time has come: The KPMG survey of sustainability reporting. Retrieved from https://home.kpmg

Krueger P, Sautner Z, Starks LT (2020) The importance of climate risks for institutional investors. Rev Finan Stud 33(3):1067–1111. https://doi.org/10.1093/rfs/hhz137

Lamberton G (2005) Sustainability accounting—a brief history and conceptual framework. Account Forum 29(1):7–26. https://doi.org/10.1016/j.accfor.2004.11.001

Lazard (2020) Sustainable investing: from niche to mainstream. Retrieved from https://www.lazard.com
Le Blanc D (2015) Towards integration at last? The sustainable development goals as a network of targets. Sustain Dev 23(3):176–187. https://doi.org/10.1002/sd.1582
Leventis S, Hasan I, Dedoulis E (2013) The cost of sin: the effect of social norms on audit pricing. Int Rev Financ Anal 29:152–165. https://doi.org/10.1016/j.irfa.2013.05.001
LMA (2019) Sustainability linked loan principles. Loan Market Association. Retrieved from https://www.lma.eu.com
Moody's Investors Service (2020) ESG: the rise of sustainable finance. Retrieved from https://www.moodys.com
Morgan JP (2020) ESG and impact investing: the new mainstream. Retrieved from https://www.jpmorgan.com
OECD (2017) Investing in climate, investing in growth. Organisation for Economic Co-operation and Development. Retrieved from https://www.oecd.org
Principles for Responsible Investment (PRI) (2018) How ESG engagement creates value for investors and companies. Retrieved from https://www.unpri.org
PwC (2019) The ESG evolution: ESG—the new normal for private equity? Retrieved from https://www.pwc.com
S&P Global (2020) The ESG advantage: exploring links to corporate financial performance. Retrieved from https://www.spglobal.com
Refinitiv (2020) Financing sustainable development: the rise of ESG in capital markets. Retrieved from https://www.refinitiv.com
Robeco (2019) Sustainable investing: the next decade. Retrieved from https://www.robeco.com
Sachs JD (2012) From millennium development goals to sustainable development goals. The Lancet 379(9832):2206–2211. https://doi.org/10.1016/S0140-6736(12)60685-0
Solar Sister (2018) Annual Report. Retrieved from https://www.solarsister.org
TCFD (2017) Recommendations of the task force on climate-related financial disclosures. Task force on climate-related financial disclosures. Retrieved from https://www.fsb-tcfd.org
TCFD (2018) TCFD 2018 status report. Retrieved from https://www.fsb-tcfd.org
UNCTAD (2014) World Investment Report 2014: investing in the SDGs. United Nations conference on trade and development. Retrieved from https://www.unctad.org
UNDP (2016) Financing the 2030 agenda: an introductory guidebook for UNDP country offices. United Nations Development Programme. Retrieved from https://www.undp.org
UNEP (2019) Global environment outlook—GEO-6: healthy planet, healthy people. United Nations Environment Programme. Retrieved from https://www.unenvironment.org
UNEP Finance Initiative (2018) Rethinking impact to finance the SDGs. Retrieved from https://www.unepfi.org
UNEP Finance Initiative (2019) Principles for responsible banking. Retrieved from https://www.unepfi.org
United Nations Global Compact (2018) The ten principles of the UN Global Compact. Retrieved from https://www.unglobalcompact.org
United Nations (2015) Transforming our world: the 2030 agenda for sustainable development. Retrieved from https://www.un.org
United Nations (2015) Addis Ababa action agenda of the third international conference on financing for development. Retrieved from https://www.un.org
United Nations (2018) Financing for development: progress and prospects 2018. Retrieved from https://www.un.org
WBCSD (2018) SDG Business Hub. World Business Council for Sustainable Development. Retrieved from https://www.wbcsd.org
WEF (2020) Unlocking technology for the global goals. World Economic Forum. Retrieved from https://www.weforum.org
World Bank (2016) Financing sustainable development: ideas for action 2016. Retrieved from https://www.worldbank.org

World Bank (2016) Public-private partnerships reference guide. Retrieved from https://www.worldbank.org

World Economic Forum (2018) The Global Risks Report 2018. Retrieved from https://www.weforum.org

World Resources Institute (WRI) (2019) The business case for sustainable finance. Retrieved from https://www.wri.org

WWF (2017) Amazon rainforest conservation. World Wide Fund for Nature. Retrieved from https://www.worldwildlife.org

WWF (2017) Sustainable development goals and WWF. World Wide Fund for Nature. Retrieved from https://www.worldwildlife.org

Yunus M (2007) Creating a world without poverty: social business and the future of capitalism. PublicAffairs

Chapter 9
The Role of Carbon Credits in Promoting Sustainable Energy Through Sustainable Design and Technologies

Monu Bhardwaj, Namrata Prakash, and Rupa Khanna Mahlotra

Abstract The integration of carbon credits into sustainable design represents a pivotal approach to mitigating climate change by incentivizing environmentally responsible practices. This research explores the role of carbon credits within the framework of sustainable design, focusing on how their inclusion can drive innovation, reduce carbon footprints, and promote eco-friendly building solutions. By aligning financial incentives with sustainability goals, carbon credits encourage architects, engineers, and developers to adopt energy-efficient technologies, low-emission materials, and green construction methodologies. The study investigates the potential of carbon credits in optimizing resource efficiency, reducing lifecycle emissions, and fostering circular economy principles in design processes. A comprehensive analysis of case studies, market mechanisms, and regulatory policies will be conducted to assess the feasibility and long-term impact of carbon credit integration in various sectors, including architecture, urban planning, and industrial design. The findings aim to provide strategic insights for stakeholders seeking to align environmental and economic goals through innovative, sustainable design practices.

Keywords Carbon credits · Sustainable design · Carbon footprint · Green building · Circular economy · Energy efficiency · Lifecycle emissions · Regulatory policies

9.1 Introduction

It is frequently asserted that energy is essential for growth and progress. The transfer of energy across varying forms is what fuels daily human lives. As civilization has evolved, the earth's fossil fuel energy resources have permitted greater developments

M. Bhardwaj (✉) · N. Prakash
School of Management, Graphic Era Hill University, Dehradun, India
e-mail: monubhardwaj865@gmail.com

R. K. Mahlotra
School of Commerce, Graphic Era Deemed to be University, Dehradun, India

R. Kumar (ed.), *Sustainable Finance*, Sustainability Solutions,
https://doi.org/10.1007/978-3-031-97962-0_9

in livelihood, manufacturing, prosperity, medicine, and trade. However, the perils of greed, wastage, and pollution have started receding some of the benefits of fossil fuel energy resources. There is an urgent need for the eradication of the disastrous effects of these anthropogenic emissions promptly. As hopes provide width for advancement in future activities, innovative energy sources promise bright and cleanliness for upcoming generations. Creativity usually broadens the horizon of modernity, thereby developing civilization. Still, many believe improving and designing technological developments will do more harm than good, further hurting the earth's environment and stability.

Every day the world wakes with many routines, which prompt the value chains of world industries running systematically. Manufacturing intentionally or unintentionally leads to greenhouse gas emissions, such as carbon dioxide and methane (Chakraborty et al. 2024). This has, in turn, changed the earth's average surface temperature and glacial balance, creating other weather extremes (Awasthi et al. 2023). It is important to preserve the natural balance of the environment as everyone is an integral part or component of it. An understanding of the aspects of everyday life as cause and effect is essential. Environmental sustainability has become a foremost objective globally and emphasizes numerous endeavors. Owing to the greater need, sustainability led to the invention and integration of energy-efficient-based services and technologies, namely sustainable design and sustainable design technologies, into the global competitive market.

The footprints of industries can never be wiped out but can be tended to develop an accountable picture by adopting and implementing sustainable design and sustainable design technologies. They assist in ensuring responsible and obligatory actions toward carbon footprints. Due to many environmental regulations regarding carbon emissions, investing in sustainable technologies minimizes emission penalties. Consequently, a systematic desire to develop sustainable designs within manufacturing industries has emerged. Sustainable design was developed in the 1990s (Kumar et al. 2024a, b). It is often thought of as a development process or blueprint of industrial operations involving materials, methods, outputs, and disposal of outputs to efficiently use energy resources and create less adjustment to the natural environment. Sustainable design technologies include the optimization of designs, methods, and operation simulations based on the mass and energy balance of a system (Kumar et al. 2023a, b, c, d, e, f).

9.2 Understanding Carbon Credits

The necessity to tackle climate change and its adverse effects has never been more pressing. In this context, carbon credits have emerged as a critical market-based approach to tackling greenhouse gas (GHG) emissions. They play a pivotal role in promoting sustainable energy through sustainable designs and technologies (Kumar et al. 2024a, b).

Carbon credits represent a validity of 1 ton of carbon dioxide removed, reduced, or sequestered from the atmosphere. This can be achieved through various means, such as carbon offsetting initiatives, renewable energy efforts, and reforestation projects. The cost of carbon credits varies depending on the initiative and engaged parties in the carbon credit trading market. This trade on the right to pollute has undergone several ups and downs.

In some western countries, carbon trading has become compulsory, with regulated carbon credits purchased from companies classified as carbon emitters. Such market-based approaches motivate organizations to create and invest in projects directed toward reducing pollution. For companies with contaminated projects at the ground level, purchasing carbon credits is often seen as a cheaper option than relocating projects or investing funds to recover these areas. In developing countries, however, companies interested in investing in carbon credits are compelled to fulfill a comprehensive series of complex preconditions to verify the validity of the carbon credits, the project affects the region and pollutes in a legitimate manner while not allowing shady easier carbon projects (Kumar et al. 2024a, b).

9.2.1 Definition and Concept

In recent years, the topic of global warming has grown increasingly popular, prompting many to explore ways of allowing for a sustainable future. The issue arose from the discovery of the ozone layer, which protects the Earth from harmful ultraviolet rays, and how it's being depleted by greenhouse gas emissions. These emissions are a result of the burning of fossil fuels for energy and other resources. In order to allow for the continued economic growth of countries, it was established that companies could trade greenhouse gas (GHG) emissions by capturing as much as they emit. The idea of carbon credits emerged as such a tradable commodity.

Organizations such as the Intergovernmental Panel on Climate Change (IPCC) arose to combat these emissions and to inform humanity of the necessity to make changes. In response to the disasters rampant in the world and the ever-present threat of global warming and climate change, these organizations created international treaties to be signed by countries globally vowing to increase the understanding of climate change, its impact if not mitigated, and taking action to ensure it is stifled. Historically, the most impactful of these treaties was the Kyoto Protocol, which was established in 1992 and resulted in GHG emission reduction targets by 2008–2012 for developed countries.

On a global scale, a limit was established for the amount of pollution that could be emitted by developing countries over the years 2008–2012. However, in the United States, it was recognized that the market economy would cause these countries to seek the most economically viable avenues to meet these limitations, which would traditionally involve more environmentally detrimental routes. To prevent achieving these targets along less desirable means, a method of allowing companies to trade their emissions and thus allowing them to further industrialize to a sustainable degree was rendered necessary. In this way, it would be economically prohibitive to pursue harmful activities while also profitably pursued. A trading market emerged with maximum limits that could be lowered as countries promoted sustainable intentions.

9.2.2 *Types of Carbon Credits*

Carbon credits can be classified into two main categories: compliance carbon credits and voluntary carbon credits. Compliance carbon credits are created through government-mandated regulatory programs or schemes to offset carbon emissions. An example of compliance carbon credits is the European Union Emissions Trading System (EU ETS), which was launched in 2005 as a "cap and trade" mechanism. The cap and trade mechanism requires companies to cap their total carbon emissions. They are allocated a certain number of carbon credits based on emissions benchmarks. If they emit less than their allocations, they receive surplus credits; if they emit more, they must purchase equal amounts of credits. Under the trading scheme, credits can be exchanged in an organized marketplace. The EU emissions trading scheme is the world's largest compliance carbon credit market, encompassing the 27 member states, Norway, Iceland, and Liechtenstein (Li et al. 2022).

In 2005, the Canadian province of Alberta was the first jurisdiction outside of Europe to create a compliance carbon credit regime. Under the Alberta Climate Change and Emissions Management Act, large emitters were required to limit emissions via a "best-in-class" approach based on industry benchmarks. Further regulations were phased in to reduce total emissions by 50% industry-wide from 2006 levels by 2021. As an alternative to meeting benchmark compliance, companies can purchase offset credits from emission reduction projects. In this system, electricity generation covered by compliance credits met industry benchmarks and carbon neutral credit status. Burnt oil sands emissions credits were equal to double their volume.

Voluntary carbon credits, on the other hand, are created under voluntary carbon credit project protocols and standards and are used to offset emissions voluntarily. Voluntary credits can be sold in compliance markets if they meet specific requirements. In Canada, the Calgary-based non-profit organization Offsetters Climate Neutral Society was the first carbon credit retailer to broker a voluntary credit in a compliance market ground-breaking sale. It sold carbon credits capped under the BC Climate Action Secretariat assisted Carbon Neutral Government Initiative to the City of Calgary in 2010, making it eligible under Alberta's compliance markets. This project generated credits for use both within Alberta and in compliance markets; credits cap and trade models sold outside Alberta compliance markets maintain price stability, but cannot be sold agency credits under compliance market regulations (Kreibich and Hermwille 2021).

9.3 The Importance of Sustainable Energy and Design

Energy is vital for development and growth of nations. Energy production through fossil fuels such as coal, oil and natural gas has resulted in huge economic growth but has brought with it a huge toll on the environment and the nations' health. Smog resulting from vehicular and industrial emissions chokes cities and has caused millions of deaths and hospitalizations over the decades. Other hazardous particulate emissions from energy burning hastened the premature decline of the global ozone layer. Finally, the burning of fossil fuel releases greenhouse gases due to which the global temperature is steadily rising resulting in catastrophic changes in climate and weather patterns.

Energy is the bedrock of modern economies. Nations are dependent on energy investments to drive their economies' growth. However, the historical overdependence on fossil fuel burning has crippled the economy and caused irreversible damage to the ecosystems. The overcatch of fossil fuels has also cost nations their lucrative economic growth. So, sustainable energy investment is now the foremost need of the hour.

Energy derived from renewable sources such as solar, wind, hydro, tide, geothermal, biomass, etc., is touted to be the most sustainable and environmentally friendly option. The major limitation of this energy is that its production is highly dependent on the god gifted natural resources and hence huge research works are being carried out to convert other wastes into energy. Other energy producing technologies like energy from methane, biodiesel, hydrogen, fusion, geothermal energy and bio-digesting have taken concept shape and are being explored further. Use of phospholipid and sugar fatty acids as an alternative to fossil fuel for energy storage and transformation is a promising area for future research (Kamal et al. 2022).

Sustainable design refers to the development of such technologies and design of structures which follow the principles of sustainable development. It helps to promote the conservation of precious non-renewable resources and other environmentally friendly technologies as well as social, economic and aesthetic considerations. It aims to lessen the negative impact of human activities on various ecosystems by reversibly utilizing the resources through innovative designs. Sustainable design is very vast and a lot of areas are yet to explore. But technologies are there in computers, electronics, automobiles, machinery, sensors, transportation, biomedical, structures and buildings, packaging, etc. which claim to be sustainable (Khan et al. 2023).

9.3.1 Environmental Benefits

Amidst the escalating concerns of climate change and its adverse impacts on biodiversity and human livelihoods, the growing adoption of sustainable design and technology has emerged as a viable solution. The expansion of carbon credit markets presents a unique opportunity to tackle climate change while promoting sustainable design and technology. The World Bank estimates that 100 carbon credit markets exist worldwide, and over four billion tons of carbon credits are issued annually. With each carbon credit representing the elimination of one ton of CO_2 from the atmosphere, the total monetary worth of the 100 markets is around 60 billion dollars. The worth of a single carbon credit is approximately 15–20 dollars. Hence, capitalized design projects, such as the construction of villas, commercial buildings, or offices, can receive carbon credits from the government for doing so in a sustainable way. These credits can then be traded in the carbon credit market. Since it takes a team of civil engineers and architects approximately three years to complete a design project, an estimated four to eight million projects capitalized by architects can be expected to be completed in the succeeding decades. With each project earning an average of 1000 carbon credits, the potential profits for architects over the decades could reach trillions of dollars (Kim and Lee 2021).

The goal is to convince developers to carry out sustainable design and energy projects. However, the real responsibility for tackling the impact of climate change lies with the government through the establishment of appropriate rules and regulations, not with designers and specific energy projects. Setting up rules for a carbon credit market is absolutely necessary. Taking the case of India, the commitment made to the Kyoto Protocol to reduce greenhouse gas emissions can be fulfilled by setting such a market. Appropriating designs for sustainable construction or freight propulsion may not be in the best interest of the industry. However, for carbon credit, many developers and designers can be expected to race toward sustainable projects, making investments in such technology and design. Designers who worked with alternative building materials and energy sources have an added advantage due to their know-how and experience.

9.3.2 Economic Advantages

Sustainable energy design and technologies encompass a range of strategies intended to reduce energy usage and ultimately the impact on the natural environment. There are a number of potential advantages to sustainable energy generation including: reduction in environmental impact; security of energy supply; protection against potentially rising energy prices; cost advantages due to the reduced risk in energy expenditure. Many of these benefits also apply to energy efficiency and demand side management technologies (Østergaard et al. 2020).

Governments, companies and individuals concerned about the impact of climate change are increasingly looking at measures they can take to address the potential danger. For many countries and industries this has resulted in the recognition that the security of the continued use of fossil fuels is becoming increasingly tenuous. There are signs that at current extraction rates the world's known reserves, most located in turbulent political areas, will be exhausted in the not too distant future. Attempts to find or stimulate the creation of alternatives, for example hydrogen and biofuels, are far from certain. Similarly attempts to control and contain the use of fossil fuels, for example the use of carbon taxes, may not discourage countries particularly driven by economic growth, for example India and China, from pursuing environmentally damaging energy policies, leading to widespread climate change.

One potential approach is to promote the use of energy efficiency and demand side management technologies. Early designs such as pneumatic tubes, mechanical buttons and clocks however would be entirely foreign to today's users. With calls for more interactive devices greater consideration is needed of how emerging smart technologies will be used and needed on a day-to-day basis. Technologies such as energy meters and devices designed to turn off unused appliances may presently challenge a user's culture and need to develop new negotiating skills. Alternatively the promise of smart technologies may appear as a vulnerable security due to the fear of revealing too much information about a household, simply a lot of work to interpret and filter out unwanted information overload or a bureaucratic inconvenience. In addition it is necessary to consider how new appliances may impact on work life balance and notions of escapism. The hope is that using such insights should help guide technological design in a manner that would enable positive future scenarios while avoiding critical problematic pitfalls (Groppi et al. 2021).

9.4 Technologies for Sustainable Energy

In the continuous pursuit of sustainable development, nations around the planet have come together to mitigate climate change and contribute toward the decarbonization of global economies. Carbon credits have proven to be a very versatile tool in achieving environmentally friendly goals. With time, carbon credits have evolved to help fund the renewables' energy transition and at the same time actively make sure the funds are used in a sustainable manner through sustainable design and technologies. This essay will discuss how carbon credits can be utilized on a global scale to promote the development of sustainable energy through the use of sustainable technologies and sustainable design.

Due to the alarming increase in fossil fuel usage in recent years, new sources for energy generation need to be actively considered. Renewable energy has gained a lot of traction in the recent decades to fulfill this need. However, renewable energy sources are not only limited to solar energy, wind energy, and hydropower, but countries are actively working on new technologies and design approaches to harness the right conditions for alternative energy generation methods.

Sun is the most abundant source of energy in this planet, and each country receives different amounts of solar radiation based on its geographical location. Solar energy can be harnessed and converted into heat or electricity. The methods that are commonly employed to generate solar energy include solar thermal systems and solar photovoltaic systems. Solar thermal systems utilize solar collectors to absorb the solar radiation and convert it into heat. The method is mainly used for producing steam, which is used to power turbine generators. On the other hand, solar photovoltaic systems employ photovoltaic cells to convert solar radiation into electricity. Flat plate, parabolic trough, central receiver, and solar chimney are most commonly used solar thermal systems, while monocrystalline, polycrystalline, and thin film are widely used photovoltaic systems. Countries with more than 10% of their total installed capacity from solar energy sources include Germany, Spain, Italy, Japan, Florida, China, India, USA, and many others. China is widely regarded as the world's solar power champion with its strong support policies, large and sufficient investments and financial aids (Adewumi et al. 2024).

The wind is a naturally available source of energy that can be converted into electricity using wind turbines. Wind energy is available offshore and onshore and is harnessed based on the economic conditions and site locations. The vast majority of the installed wind energy plants in the word are considered onshore, however, due to the high land occupation, operational and social costs, environmental impacts, emergence of more innovations and technologies such as floating wind farms, offshore applications have gained a lot of traction and considerations. The most widespread method to harness the wind as a renewable source of energy is the horizontal axis wind turbine, while vertical axis wind turbine technologies are still gaining attention in recent years for small-scale applications. Countries with more than 10% of their total installed capacity from wind energy sources include Denmark, Spain, Portugal, Ireland, Germany, UK, USA, Canada, India, China, and many others. The vast majority of the wind power generation in the world is made in Denmark, with around 25%.

The natural flow of water is a globally available source of energy that can be converted into electricity through a number of different technologies. The energy of rivers and different water streams can be harnessed by installing dams and using turbines and generators in a similar fashion adopted by fossil fuel power generation plants (Sharma et al. 2024). Unlike conventional energy generation methods, the dams can be run on base load and can ensure a high efficiency rate of above 90% throughout the year. Different technologies have also been developed to use the tides and currents of the oceans to generate energy. The technology to harness sea currents and waves is still under development, but it has emerged as a very promising approach to sustainable energy generation due to the predictability of these resources. Countries with more than 10% of their total installed capacity from hydropower generation include Canada, Brazil, Norway, Venezuela, New Zealand, Colombia, Thailand, Sweden, India, China, and many others.

9.4.1 Solar Energy

Solar energy, harnessed from the sun's rays, stands as one of the most viable renewable energy sources. Its availability is more or less constant across the globe. In terms of technology, solar energy can be harnessed either actively or passively. Active solar energy involves the adoption of technologies that convert direct sunlight into heat energy or electricity; an example is the utilization of photovoltaic cells. On the other hand, passive solar energy involves the adoption of technologies that accumulate, store, and distribute solar energy without converting it into heat or electricity; an example is the utilization of direct sunlight to keep buildings warm. Solar energy is applicable at almost every stage of energy production and use (Kumar et al. 2023a, b, c, d, e, f).

The development and commercialization of solar energy technologies have been promoted through various governmental and non-governmental programs across the globe. In view of the limited storage and adverse environmental effects associated with fossil fuels, many developing countries like India have launched solar energy technology programs to reduce dependence on fossil fuels in the near future.

There are many prospects associated with the usage of solar energy technologies. Countries with otherwise adverse climatic conditions, such as landlocked regions and polar regions, can still harness solar energy. In times to come, the advancement of technologies would increase the conversion efficiency drastically. There is significant potential for the application of solar energy technologies, not just for power generation but also for industries that involve direct heating of components. However, a large percentage of solar technologies remain untapped (Jäger-Waldau 2020).

Increasing awareness of the harmful consequences of fossil fuels vis-à-vis carbon credit schemes has spurred the growth of more non-conventional sources of energy, largescale study of the feasibility of alternative technologies, and development of scientific models/systems for efficiency assessment and optimization of design characteristics. Solar energy with its near constant availability across the globe—particularly in tropical countries—represents one of the most viable renewable sources of energy.

9.4.2 Wind Power

Wind power has a long history; through antiquity, people harnessed wind energy. The earliest known sails were found on boats in Egypt c. 3000 BCE, and the earliest windmill archeological evidence dates to 500–900 CE in Persia. This technology quickly spread to Europe, where the windmill designs evolved. During the industrial revolution in the eighteenth century, wind energy was also applied to water pumping. But it ultimately lost the competition against fossil fuels and was nearly abandoned until the high oil prices of the 1970s sparked interest in alternative energy sources, especially among Scandinavian countries. The Delft University of Technology in the Netherlands constructed the first large modern wind turbine in 1957, whose design became the foundation for turbines constructed later in the 1960s in Denmark and California. As these new turbines proved successful, interest in wind energy grew rapidly in many countries as part of the effort to bring down the cost of energy (Kreps 2020).

Wind power uses the kinetic energy of moving air masses to turn the blades of a turbine mounted on a tower high above the ground. Wind blows due to pressure differences in the earth's atmosphere caused by the sun's heat. The larger the temperature differences, the stronger the winds. The wind speed varies with time as well as with height and location. This variability is both a challenge and an opportunity. In some places, for example, the North Sea, the population density is high, and this is where most of the energy demand is. But this is also where the winds are the most variable, making it more difficult to forecast. Therefore, there is a strong interest in predicting how the wind speed will develop in time and space.

Wind energy has several advantages over traditional energy resources. First, it is environmentally friendly and does not pollute. Second, wind energy is inexhaustible, as long as the sun shines and the Earth turns. Wind farms also create job opportunities. Wind power is already a growing source of energy worldwide. By the end of 2005, the total installed capacity of wind power around the world exceeded the 65 GW mark and was power generating capacity equating to a complete electrical system. Developers, owners or operators of wind projects can claim carbon credits to offset their greenhouse gas (GHG) emissions impact, in part promoting projects focused on renewable energy sources, which are usually low GHG emitters than fossil fuel-based generation.

9.4.3 Hydropower

Hydropower has long been one of the largest renewable energy sources available. Its large-scale implementation is capable of meeting increased energy needs while simultaneously reducing carbon emissions. Hydropower has power potentially large enough to meet the entire energy needs of the United States. New hydropower is one of the most cost-effective large-scale energy technologies, and existing dams may still accommodate additional hydropower. Hydropower requires careful consideration to maximize the beneficial impact of construction and operation. Some dams are geographically undesirable, while others may impede fish migration, require extreme water level fluctuation, or create an electrolyte lake effect. Inadequate consideration may lead to large catastrophic events. Flooding or staggering loss of hydropower may affect prices, cut power or cause large cascading events across the grid. Nevertheless, properly siting, designing, and operating hydropower plants, using the most modern technology and techniques, can minimize challenges and maximize benefits (Siri et al. 2021).

Ongoing climate change is likely to stress hydropower resources of plants that rely on snowmelt or cold-water fish. With changes in precipitation and temperature, the capacity reliability of current hydrant dams may change, potentially leading to large shifts in installed capacity reliability and displaced generation such as coal-fired combustion. Hydropower's existing plants account for 95% of its generation. Plants generating naturally with no dam reservoirs would be the least able to respond. Those with pumped hydro might be able to respond by increasing storage, but reliability would still be decreased in general in the further future. Since 2001, approximately 88% of new hydropower generation capacity in the U.S. has been through improving the efficiency of existing facilities.

Further construction of dams should consider downstream impacts, such as delta and estuary land loss mostly to subsidence and sediment trapping. While hydropower may contribute to mitigation efforts, care should be taken for local environmental impacts, which are often significant. During this past century, carbon pollution from fossil fuel power plants has substantially altered the planet's climate. Meanwhile, new efforts in electricity generation have not resulted in meaningful cuts in carbon emissions emissions, to which the dirty fossil fuel plants still contribute over 66% of global electric generation energy needs. The vast scale of electricity generation from hydropower plants needs to be implemented in order to have a meaningful impact. No other energy source has the capacity to come close to meeting current power needs 100-fold and avoid catastrophic climate change.

9.4.4 Geothermal Energy

Geothermal energy captures the heat from the middle of the earth, its aim is to convert thermal energy for its use to produce electricity or provide heating. Considered a clean renewable technology, it has been used traditionally in China, Mexico and Japan for bathing, heating and cooking, covering all needs in water, sanitation and workplace heating. Nowadays, modern technology has taken control over this resource, with utilites providing electricity and heating through power plants. There are five main applications for geothermal energy: to provide electricity and heating for small and big buildings (direct use), district heating (heating neighborhoods), electricity production and co-generation (electricity and district heating) (Tester et al. 2021).

There are three main types of geothermal power plants: single flash, double flash and binary cycle plants. America is the world leader in geothermal applied capacity and is setting an example for the rest of the world. Fossil fuels have been the main source of energy since they were discovered, but they are non-renewable resources that will eventually deplete with use. In 1955, scientists and government agencies began interesting in alternative resources of energy with environmental and economic impacts. Since then, many technologies converting natural energy have been studied, analyzed and developed in prototype and demonstration facilities. Hydropower was the first renewable resource to be used, followed shortly by wind energy and geothermal. Today, photovoltaic solar panels are starting to be utilized with commercial success.

Geothermal energy is abundant, estimated to provide around 344 million terajoules of heat and 33,000 million gigajoules of energy per year. Currently, around 2300 Mw of geothermal based electricity is being used globally, being nonpolluting and having beneficial impacts on economic activities and quality of living. Technologies for geothermal energy extraction, conversion and distribution are not well known, and its environmental impacts are mostly unknown or neglected. Creating international or cross-border business in geothermal energy could make a difference.

Interest in geothermal energy has grown recently with economic changes, civil wars and military conflicts in many countries.

9.5 Benefits of Utilizing Carbon Credits

Carbon credits have emerged as a pivotal tool in the broader strategy to tackle climate change. By establishing a market for greenhouse gas (GHG) emissions through a cap-and-trade system, carbon credits create an economic incentive for organizations and nations to reduce their emissions. These credits can be traded within this system, with each credit representing a reduction of one metric ton of carbon dioxide emissions or its equivalent in other GHGs. The expansion of supply and demand for carbon credits is expected to lead to significant investments in sustainable technologies.

While it is vital to recognize the limits of carbon credits in effectively dealing with climate change, they can provide additional, valuable help. Nations with high marginal costs of abating emissions will find new trades an attractive and low-cost alternative to traditional emissions reductions strategies. The market for carbon credits enables these nations to fund GHG reductions elsewhere. On the other hand, nations with low marginal costs of reducing emissions can lower the abatement costs of industrialized nations through new trades. In other words, these credits provide the framework to deal with an artificially constructed "global commons" problem through an efficient market mechanism.

To further elucidate the rationale behind carbon credits, it is useful to investigate the complexities surrounding the concept of property rights. Monarchies, military victors, and acts of disposal by powerful political entities granted property rights over land to individuals in early human history. Although this method was effective in establishing dominion over land-based resources, property rights over non-tangible resources such as GHGs are both more complex and equally critical. Over the last century, substantial advancements in technology have allowed the exploitation of resources that were previously unconceivable. The atmosphere was falsely viewed as infinite by humanity, and an economy that increases dematerialization of all activities was believed to be a safeguard for the biosphere. However, mathematical models predicted the emission of carbon dioxide to exceed the highest natural concentration of 280 ppmv for two million years. These predictions became a reality. The attainment of efficiency and wealth was falsely interpreted as non-conflicting objectives (Chandra et al. 2023).

9.6 Integration of Carbon Credits in Sustainable Design

The successful integration of carbon credits in the design process depends heavily on the availability of accurate scenarios, predictions associated with modeling strategies, the willingness of actors to engage in carbon management schemes, clear regulatory definitions of unitary costs and benefits, as well as the temporal scale adopted—that is, whether the perspective is, for example, a short design cycle or several generations of building life. Examples exist in varying degrees of detail where aspects of climate change have been integrated in the design process, ranging from abatement strategies to the effects of climate change, from creek diversions to bioclimatic design. Two case studies illustrate this integration, one involving a commercial building (Sustainable Strategies: A European Award Winning Building), while the other is set in a semi-arid environment (Palm Springs Visitor Center) (Ahmadi et al. 2020).

In the first case study, EnergyPlus simulations with a greenhouse model were used to explore energy penalties associated with climate change referring to future changes in temperature, solar radiation, humidity, wind patterns and rainfall. This analysis was further elaborated through modeling the building's future development under two climate scenarios: the 'climate best' (today's) and 'climate worst' (25% increase in temperature, 25% reduction in groundwater, increased cloud cover). Results show that if no action is taken, temperature and energy penetration will almost double in 2051 as compared to today's situation. However, by considering measures derived from a broad range of design strategies, the adverse effect of climate change can be reduced by almost two-thirds. Interestingly, it was further found that good design is not necessarily energy emergency-proof. Some passive strategies that remain neutral under current climate conditions may have negative effects in future climates (for instance thick northwards blinds). Resilient strategies include water/temperature buffering systems that have activity networks transcending outside, variable thermic mass levels, as well as building envelopes with variable U-values (Usta et al. 2021).

9.6.1 Case Studies

Incorporating carbon credits into sustainable design through successful case studies. Researching and learning from specific instances where these concepts were put into practice can provide valuable insights. Exploring examples of buildings or designs that successfully included sustainable design practices and obtained carbon credit certification, along with the role of design thinking in these efforts, showcasing how research, creativity, and a human-centered approach were utilized.

The Bullitt Center, located in Seattle, Washington, is widely recognized as one of the greenest commercial buildings in the world and serves as a successful case study in sustainable design. Completed in 2013, the Bullitt Center was designed to meet the Living Building Challenge, a rigorous green building certification program that requires buildings to produce more energy than they consume, capture and treat all water on-site, and operate without the use of toxic materials. The building employs a range of sustainable design features, including a green roof, rainwater harvesting system, and extensive use of natural daylight to reduce the need for artificial lighting. One of the unique aspects of the Bullitt Center is its use of carbon credits to offset its greenhouse gas emissions. The building participated in the U.S.

Environmental Protection Agency's Green Power Partnership, which enables organizations to purchase renewable energy credits (RECs) to support renewable energy production and offset their emissions. The Bullitt Center's commitment to sustainability and its use of carbon credits has positioned the building as a model for future development.

The Bosco Verticale (Vertical Forest) is a pair of residential towers located in Milan, Italy, widely recognized as an innovative case study in sustainable design. Completed in 2014, the towers are designed to accommodate a density of 9000 trees, 100,000 shrubs, and 14,000 plants over a planted area of 9000 m^2 (97,000 ft^2), the equivalent of a wooded area of 7 ha (17 acres). In addition to the cooling effects of the plants, which mitigate heat, reduce energy needs, and co2 emissions, the project increases humidity, attracts insects and birds, thereby improving biodiversity, as well as mitigates the 'urban heat island' effect in the city of Milan. The vegetation is expected to absorb 9 tons of coarse particulate matter each year, and 30 tons of fine particulate matter (PM10), and capture 47 tons of CO_2 each year while reducing energy consumption for cooling and heating by about 30% (on the north side) and 40% (on the south side) per year. The towers shade oversized windows and reduce dependence on energy-intensive air conditioning. The Vertical Forest integrates cutting-edge technology, such as monitoring systems that control temperature, humidity, illumination levels, and burning substances in grilling zones, and detect seismic activity for a possible alert to residents. The project won several awards, including the International Highrise Award 2014 and 2015 Best Tall Building Worldwide from the Council on Tall Buildings and Urban Habitat. Both case studies serve as successful examples of sustainable design that could teach lessons for future projects, addressing questions on what aspects of the project might be replicated elsewhere and what role design thinking played in the project.

9.7 Challenges and Opportunities

As the world moves toward a low-carbon economy, carbon credits are fast becoming an essential tool in the fight against climate change. By providing financial incentives for the reduction of greenhouse gas emissions, carbon credits have the potential to accelerate the establishment of a more sustainable energy economy. However, the wide-scale implementation of carbon credits will not be without its challenges. Most importantly, there will need to be a confluence of interest between governments, institutions, corporations, small and medium businesses, and civil societies.

Barriers to the implementation of carbon credits include: lack of political commitment; nationally-driven standards and rules that do not provide a level playing field; lack of regulatory or legal framework to govern carbon credit systems; systemic weaknesses of global credit markets; lack of integrity of carbon credits and carbon credit schemes; lack of capacity to assess the real net gain/benefit of projects; inability to create sustainable business models; and uncertainties in project development and implementation. To address most of these barriers, stakeholders are urged to cooperate strategically and realistically to develop carbon credit systems that are based on internationally agreed systems, and tradeable on a robust global credit market. As a basis to cooperate, each carbon credit scheme should have systems that guarantee quality and sustainability, address the developmental needs of the poor, and promote the replication of good projects and activities. That businesses and technologies are also not subject to distortion and rigging is very important. An institution similar to the World Trade Organisation but focused on carbon credit trade is proposed.

Although there are significant obstacles, pathways and opportunities do exist for the effective implementation of carbon credits. Applying a precautionary, judicious approach based on experience is typically advisable when starting a new venture, especially one as politically sensitive as carbon credits. Simple, creative and transparent attempts should nevertheless be made to generate interest and excitement in carbon credits, because at present there is a great deal of apathy among stakeholders in the global South. Research is also needed to enhance understanding of the role carbon credit systems can play both in mitigating climate change and improving social equity and environmental conditions.

9.7.1 Barriers to Implementation

The promotion of sustainability by carbon market mechanisms generates opportunities for greater application of sustainable design principles in new projects and for sustainable reconstruction in existing properties, including those that have been historically significant. However, there are barriers to further implementation of carbon credits as an economic incentive for use of sustainable design technologies and practices throughout project design and construction. Three major concerns for practical implementation surfaced in discussions with professional peers in related fields: capital cost, economic efficiency, and owner involvement in committed sustainable design. There are three barriers to implementation: equity and impact on low-income communities, currency and market volatility, and current emphasis on timber carbon credits. On the notions of equity and impact on low-income communities, many community professionals are concerned that carbon credit transactions may exacerbate the division between the rich and poor. Community engagement and participation in carbon sequestration and crediting must be enhanced in proposed market use of forests both nationally and internationally. While there is increasing concern about ownership and rights of indigenous people to forests in their lands currently under corporate or government control, the larger picture of carbon credits is clearly one of commoditization, centralization, and consolidation of power. On the issue of currency and market volatility, since there is currently an overbalance between credits and debits in the voluntary market, the concern of many practitioners is whether market prices may fall so low that it will be uneconomic to have properties designed and certified to be sustainable. Many designers are enthusiastic about carbon credits and credits in general becoming a currency in the marketplace, which will be exciting, but not without many potential problems and pitfalls. On the current emphasis on timber carbon credits, there is concern, particularly among ecologists, that there may be an overemphasis in contending with the threat of global warming, in terms of timber carbon credits, to the exclusion of other functions and services that forests provide, like wildlife habitat, biodiversity, clean air and water, social and cultural values, etc.

9.7.2 Potential Solutions

Over the years, the realization that drastic measures needed to be taken against anthropogenic climate change has prevailed. Since the Industrial Revolution, there has been an alarming increase in man-made greenhouse gas emissions in the atmosphere due to extensive coal usage and the exploitation of fossil fuels in general, resulting in a net increase in warming effect. Climate change leads to deteriorating environmental and social conditions on Earth, so much so that the future of life on Earth as we know it is severely threatened. In the last few decades, countries have been more aware and cooperative to combat climate change on a regional, national, and local scale. As global cooperation is essential to tackle global issues, numerous compliance and voluntary carbon markets have been established worldwide. The fundamental operation of these markets is based on carbon credits, introduced by the Kyoto Protocol, which aim to limit the amount of greenhouse gases a firm can emit by allowing firms that meet their limits to sell the unused credits they possess to firms that exceed their limits. In this manner, the hope is for firms to invest in cleaner processes and technologies, benefiting the environment while also profiting financially. However, these markets, especially compliance markets, have been subject to a plethora of challenges and criticisms, limiting their efficiency and overall success.

In order to promote the success and efficiency of carbon credits in supporting sustainable energy transition by investing in sustainable design and technologies within its innovative framework, potential solutions are suggested. Firstly, the carbon markets could expand their scopes and include new sectors. Although it is vital for emissions to be limited relatively equally to effectively help the atmosphere, in practice, this is very difficult to achieve as industries where countries or firms hold a comparative advantage mainly engage in carbon markets. This would not only benefit emission reduction, but would also harbor innovation pressures for sweatshop-like sectors toward adopting more sustainable policies and technologies, benefiting poorer countries lacking in waste process and recycling technologies. Furthermore, the confidence in carbon credits needs to be restored by tackling the project types being supported within the compliance markets. The Qualified Project Types (QPTs) need to be reviewed and limited mainly to new or emerging carbon friendly processes and technologies that tackle specific and pre-defined environmental issues in order to truly benefit the atmosphere. For instance, since small-scale hydropower plants have existed for decades and have no negative impacts, it is impossible for them to face market-entry challenges. However, co-firing situation is not only detrimental to the atmosphere, but it is also making the most eligible credits possible by the EPAP for inefficiently burning rubbish. Projects need to either work with new, innovative processes or technologies, or with well-established processes or technologies but in a developing country context.

9.8 Conclusion and Future Perspectives

The issues of climate change, urbanization, and a global population expected to increase to 9 billion by 2050 present monumental challenges. Energy-related greenhouse gas emissions must be reduced by 2.3 gigatons of carbon dioxide equivalents ($GtCO_2e$), or 90%, annually by mid-century (2012 levels) to avoid a rise in global temperature of more than 20C (UNEP 2011). Global energy demand is estimated to rise by 36% from 2008 to 2035 (IEA 2010), with the developing world accounting for the majority of this rise. Energy must be utilized more efficiently and conservation strategies implemented to ensure equitable access to affordable energy while complying with environmental priorities, particularly in developing nations. As populations grow in these nations, so too does energy consumption. New building construction has almost exclusively occurred in these areas. The Center for Sustainable Energy (CSE) in the United Kingdom (UK) has independent knowledge and working relationships with developers, financial investors, and managers of social housing association homes and privately rented homes, most of which would be regarded as substandard by UK energy efficiency requirements. This coalition has used its knowledge of energy credits, carbon offsets, and compliance methodologies implemented in the UK and California to produce a program that hopes to use energy services to mitigate poverty in the developing world (Johnston, 2010; CSE Grant Application).

Carbon credits fund programs to ensure compliance with legally binding emissions limitations mandated by legislative initiatives relating to energy production and consumption. Mandatory emissions limitations were first enacted in the UK and followed by several countries and states in the USA. The resultant projects reduce emissions from polluting sources by increasing energy efficiency or by implementing alternative sustainable energy programs and technologies. These projects receive Carbon Credits and are verified and registered by a regulatory body. The market mechanism created by these credits is traded to and between responsible parties as a means for compliance with mandates (Ashby 2005). In turn, proceeds from the credits fund implementation of additional projects, creating a self-sustaining growth cycle. Verification is a paramount concern of the compliance programs as specific methodologies are used for the initial project proposal submittal and as the basis for ensuring compliance with correct project implementation (local regulation monitoring). Grievance procedures exist for all parties if projects fail to comply with initial proposals (Talozi 2007; U.S. Government 2003).

Voluntary credits have emerged in parallel to the mandatory projects, with demand driven by concerns surrounding corporate social responsibility (CSR). Companies voluntarily go beyond compliance obligations for philanthropic reasons, and sustainability responsibly. Projects are funded by purchasing these voluntary offsets, often bundled with technology and capital to make it feasible (Zhang 2005; World Bank). Verification requires less stringent initial proposal processes to be turned over to independent project operators and technology provider partnerships (Talozi 2007). A further distinction is made between Community Projects (or 'Gold Standard' projects) that require a minimum percentage of direct benefits to the developing community, as investment return is a profit motive in voluntary projects (Burtraw 2005).

Funding NA.

Conflict of Interest: There is no conflict of interest for this research.

References

Adewumi A, Usman FO, Okoli CE, Olu-lawal KA (2024) Solar energy innovations in the USA: a comparative global review. Int J Sci Res Archive 11(1):1528–1539. ijsra.net

Ahmadi L, Kannangara M, Bensebaa F (2020) Cost-effectiveness of small scale biomass supply chain and bioenergy production systems in carbon credit markets: a life cycle perspective. Sustain Energy Technol Assess 37:100627. [HTML]

Awasthi A, Pattnayak KC, Tandon A, Sarkar A, Chakraborty M (2023) Implications of climate change on surface temperature in North Indian states: evidence from CMIP6 model ensembles. Front Environ Sci 11:1264757

Chakraborty M, Singh R, Awasthi A, Kumar V, Kumar A, Kaur S (2024) Assessment of greenhouse gas emissions in Delhi landfills: analyzing factors affecting emissions and environmental implications. J Mater Cycles Waste Manag 1–13

Groppi D, Pfeifer A, Garcia DA, Krajačić G, Duić N (2021) A review on energy storage and demand side management solutions in smart energy islands. Renew Sustain Energy Rev 135:110183. academia.edu

Jäger-Waldau A (2020) The untapped area potential for photovoltaic power in the European Union. Clean Technol. mdpi.com

Kamal MM, Asharaf I, Fernandez E (2022) Optimal renewable integrated rural energy planning for sustainable energy development. Sustain Energy Technol Assess 53:102581. [HTML]

Khan SAR, Zia-Ul-Haq HM, Ponce P, Janjua L (2023) Re-investigating the impact of non-renewable and renewable energy on environmental quality: a roadmap towards sustainable development. Resour Policy. [HTML]

Kim JW, Lee JS (2021) Greening energy finance of multilateral development banks: review of the World Bank's energy project investment (1985–2019). Energies. mdpi.com

Kreibich N, Hermwille L (2021) Caught in between: credibility and feasibility of the voluntary carbon market post-2020. Climate Policy. tandfonline.com

Kreps BH (2020) The rising costs of fossil-fuel extraction: an energy crisis that will not go away. Am J Econ Sociol. [HTML]

Kumar R, Kandpal B, Ahmad V (2023) Industrial IoT (IIOT): security threats and countermeasures. In: 2023 international conference on innovative data communication technologies and application (ICIDCA). IEEE, pp 829–833

Kumar R, Malholtra RK, Grover CN (2023) Review on artificial intelligence role in implementation of Goods and Services Tax (GST) and future scope. In: 2023 international conference on artificial intelligence and smart communication (AISC). IEEE, pp 348–351

Kumar R, Sexena A, Gehlot A (2023) Artificial intelligence in smart education and futuristic challenges. In: 2023 international conference on disruptive technologies (ICDT). IEEE, pp 432–435

Kumar R, Lande A, Kumar D, Malhotra RK, Sharma A (2023) Technology bridging-in entrepreneurs and consumers in product development. In: 2023 11th international conference on intelligent systems and embedded design (ISED). IEEE, pp 1–5

Kumar R, Saxena A, Singh R (2023) Robotic process automation bridge-in banking institute and consumers. In: 2023 international conference on disruptive technologies (ICDT). IEEE, pp 428–431

Kumar CMS, Singh S, Gupta MK, Nimdeo YM, Raushan R, Deorankar AV et al (2023) Solar energy: a promising renewable source for meeting energy demand in Indian agriculture applications. Sustain Energy Technol Assess 55:102905. [HTML]

Kumar V, Kumar R, Goel R, Singh T, Kaur MJ (2024) The evolution of digital marketing technologies: a fuzzy TOPSIS approach to selected technology in E-commerce platform. Int J Syst Assur Eng Manag 1–14

Kumar R, Goel R, Singh T, Mohanty SM, Gupta D, Alkhayyat A, Khanna R (2024) Sustainable finance factors in Indian economy: analysis on policy of climate change and energy sector. Fluctuation Noise Lett 23(2)

Li J, Yao Y, Wang X (2022) The first compliance cycle of China's National emissions trading scheme: insights and implications. Carbon Neutrality. springer.com

Østergaard PA, Duic N, Noorollahi Y, Mikulcic H, Kalogirou S (2020) Sustainable development using renewable energy technology. Renew Energy 146:2430–2437. [HTML]

Sharma M, Rawat S, Kumar D, Awasthi A, Sarkar A, Sidola A et al (2024) The state of the Yamuna River: a detailed review of water quality assessment across the entire course in India. Appl Water Sci 14(8):175

Siri R, Mondal SR, Das S (2021) Hydropower: a renewable energy resource for sustainability in terms of climate change and environmental protection. Alternative Energy Resour Way Sustain Modern Soc 93–113. academia.edu

Usta DFB, Teymouri M, Chatterjee U, Koley B (2021) Temperature projections over Iran during the twenty-first century using CMIP5 models. Model Earth Syst Environ 1–12. researchgate.net

Chapter 10
The Effect of Renewable Energy Production on Global Economy: An OLS Approach

Stutee Mohanty, Suresh Kumar Sahoo, Mohammad Ali Alhussam, and Arpita Panigrahi

Abstract The study aims to investigate the link shared by variables impacting renewable energy production and global economic growth. The investigation used a panel data model from the year 2000–2022 by integrating a multidimensional data framework through the Ordinary Least Square (OLS) technique to estimate the overall effect of the variables included in the research work. The findings presented that the growth of economy, sustainability and technology factors, patents, usage of renewable energy, and research publications assisted in enhancing renewable energy output. Wind energy production has a higher speed modification than its solar counterpart. The usage of sustainable energy sources was influenced by climate and other direct and indirect costs incurred in the process. Most of the lower and middle-income countries primarily use cheaper and stable sources of energy and consider renewable energy as a secondary support. This study takes forward the call for sustainability by linking renewable energy production and usage with economic growth. It will help achieve the goals and objectives of sustainable development.

Keywords Renewable energy · Green energy · Economic growth · Carbon emissions · Sustainability · ESG · Energy production

JEL Classifications O1 · O4 · Q43 · Q56

S. Mohanty (✉)
Birla Global University, Gothapatna, India
e-mail: stutee91@gmail.com

S. K. Sahoo · A. Panigrahi
Sri Sri University, Govindpur, India

M. A. Alhussam
Kiit University, Bhubaneswar, India

R. Kumar (ed.), *Sustainable Finance*, Sustainability Solutions,
https://doi.org/10.1007/978-3-031-97962-0_10

10.1 Introduction

Carbon emissions have drastically increased with growing usage of fossil fuel ever since the industrial revolution became a reality. This has turned into a menace in the twenty-first century as carbon emissions have now gone out of control. The Kyoto Protocol, undertaken in Japan in 1997, resolved to find solutions to global environmental issues like global warming, climate change, rise in sea level, etc. It was the first ever convention to be held on the theme of sustainability/ESG.

In 2023, the best option against fossil fuels is renewable energy. Renewable energy has been divided into three categories by the International Energy Agency (IEA), namely, hydro energy, solar and wind energy, and biomass energy. First world nations are advanced compared to second and third world nations in hydro, solar, and wind energy production and usage as it demands advanced technology and thus, is expensive. However, in the case of biomass energy, third world nations are way ahead of first world nations because primitive methods are enough for excreta/biological tissue production.

Considering the above facts, research on establishing the relationship shared by gross domestic product (GDP) and renewable energy production was pioneered (Apergis and Payne 2010). The implementation of carbon tax encouraged renewable energy production and usage and enhanced energy efficiency (Lin and Li 2011). Reduction of fossil fuel usage leads to increase in renewable energy production and usage (Withana et al. 2013; Goh 2014). Laws, sustainable investment, energy consumption, demographics of the region, and reserves are contributing factors towards renewable energy production. Lopes (2014) analysed environmental, economic, and employment factors and found that carbon taxation policies and carbon consumption were influenced by them. Kumar and Majid (2020) stimulated that GDP shares a highly positive association with renewable energy output and consumption thereby, reducing dependence on fossil fuels. Damette and Marques (2019) indicated that carbon emissions and green energy output are significantly negatively linked with each other.

As it can be inferred, prior studies on renewable energy majorly focussed on economic growth and advocated that both the variables are positively related (Mohanty et al. 2024). Carbon and environmental tax were also included in the research work and results were positive. But, the nexus shared by technology, standard of living, and environmental policies has not been addressed in depth (Kumar et al. 2023a, b, c, d, e).

This research work has a threefold objective. It studies a few selected economic factors and their impact on renewable energy output and presents a policy-making direction for the future to the administration. It investigates if the environmental policies implemented by the administration at the grassroots levels are significant on the basis of population, country, and education. It also analyses the requirement for modified speed in case of renewable energy consumption by making fixed and random effects as the premise (Kumar and Khanna 2023).

This paper has five sections. Section 10.1 includes an introduction, Sect. 10.2 explains the existing literature on the subject, Sect. 10.3 covers the methodology and research model employed in the study, Sect. 10.4 describes the findings, and Sect. 10.5 showcases the conclusion and implications.

10.2 Literature Review

Frontier curve demonstrates the production trends and explains the trade-off between economic growth and sustainability. It proves the quantity of renewable sources of energy decreases with increase in production demand. Environmental Kuznets Curve is widely followed by economists to analyse the relationship shared by economic growth and sustainability.

Pollution levels initially witness a sharp hike and then subsequently fall if the economic growth of a country is low (Stern 2004a, b). Therefore, a U-shape curve represents the pollution levels in a region (Grossman and Krueger 1995). Additionally, Beckerman (1992) concluded that as a nation becomes wealthy owing to economic prosperity, pollution levels tend to decrease and sustainability concerns are prioritised.

Stern (2004a, b) took the investigation forward by using the competition theory to study the association shared by economic development and sustainability from a global context. He suggested that sustainability is neglected by developed and developing nations, in particular, as economic development is their top priority due to global competition.

Bernauer and Koubi (2011) found that pollution and air quality control do not influence pollution levels in a country but government expenditure with respect to GDP does. Lopez and Palacios (2010) proved that government expenditure as a share of GDP is a crucial predictor of pollutant levels in Europe after relating it to carbon taxes and air pollution. Accurate distribution of government spending helps in reducing environmental damage.

Laws, sustainable investment, energy consumption, demographics of the region, and reserves are contributing factors towards renewable energy production.

Leuschner (2014) focussed on China and examined hydropower production from economic growth perspective only to conclude that both share a significantly positive association. Economic development resulted in an increase in hydropower production relative to other sources of energy.

Foreign direct investments (FDI) were negatively related to deteriorating environmental standards in native countries (Lan et al. 2012; Shahbaz et al. 2015). FDI was responsible for despicable environmental depletion globally.

In the past few years, research articles have also deeply studied the role of industrialisation in increasing carbon emissions across the globe (Zhongping et al. 2011). Manufacturing industry development led to massive air pollution and environmental degradation. Hence, both the variables shared a positive relationship with each other.

The same to be true for the South Korean manufacturing industry and pollution levels prevailing in the country (Jeong et al. 2013).

Ouyang and Lin (2015) suggested that energy efficiency of the Chinese manufacturing industry helped reduce environmental damage and pollution in China by decreasing carbon emissions in the country. Raheem and Ogebe (2017) indicated the same for African nations where the governments were focussed on enhancing energy efficiency.

Sinha and Shahbaz (2018) explained that the usage of renewable sources of energy helps reduce carbon emissions. But, it was countered by Hu et al. (2023) who believed that extensive usage of non-fossil fuels increases carbon and pollution elements in the environment. Sarkodie and Strezov (2019) proved that overall plunge in the energy usage results in enhanced carbon emissions. However, when the non-fossil energy component in the total energy consumption is enhanced, it leads to decrease in carbon emissions.

Brunnermeier and Cohen (2003) examined sustainability and technology predictors in the American manufacturing industry only to conclude that sustainability patents and environment control spending are positively related to each other. Mughal et al. (2022) studied the influence of innovation on decreasing pollution and environmental degradation and indicated that scientific patents help reduce carbon emissions through enhanced energy efficiency.

Mert et al. (2019) studied if carbon emission laws and legislations assist in improving environmental conditions in BRICS nations. They suggested that non-fossil sources of energy usage and supply decrease per capita carbon emissions.

Global climate change due to environmental degradation has resulted in a sharp decline in the living standards of people in the world (Diener 2013). Beja (2012) also concluded that hot weather and air pollution massively negatively impacted the quality of life of people in Latin America.

From the literature reviewed above, we realise that the factors like population, regional demographics, and quality of education have not been studied from an environmental context. This paper will testify the current global climate and environment and subsequently, empirically analyse non-fossil energy output and economic variables.

10.3 Data and Methodology

Table 10.1 presents the countries included in the research work by the authors.

Table 10.2 showcases the details of the variables (data unit, definition, and source) included in the study by the authors.

Table 10.1 Countries included in the study

Australia	Austria	Bulgaria	Canada
Dominican Republic	Russia	Slovenia	Estonia
Sweden	Czech Republic	Poland	USA
United Kingdom	Portugal	Luxembourg	China
Lithuania	Italy	Netherlands	New Zealand
Germany	Finland	Ukraine	India
Denmark	Latvia	Croatia	Pakistan
Kazakhstan	Portugal	Ireland	Bangladesh
Slovakia	Spain	France	Norway

Source U.S. News & World Report

Table 10.2 Variables and sources

Variables	Data unit	Definition	Source
Solar energy output production	Kilowatt-hours, million	Total solar energy	UNDATA
Wind energy output production	Kilowatt-hours, million	Total wind energy	UNDATA
GDP	10 million USD	Measurement of goods and services produced and consumed in a country	World Bank
FDI	10 million USD	Difference between FDI inflow and outflow	World Bank
R&D expenses	GDP%	% of private and public expenditure	World Bank
Government expenses	10 million USD	All the expenses incurred by Government, apart from defence expenses	World Bank
Manufacturing growth	10 Million USD	Total estimate of production in the manufacturing sector	World Bank
Non fossil fuel usage	% of total energy usage	Portion of non-fossil fuel in total energy usage	World Bank
Research articles published	Number of research articles	Research articles published in the domain of applied sciences	Scopus/ WoS
Carbon/ environment tax	USD	Tax collected by the government to limit environmental damage	OECD
Technology and innovation	Number of patents	Copyright for inventions	UNDATA
Health expenses	% of GDP	Contribution of health expenditure towards GDP	UNDATA
Usage of fossil fuel	Kilowatt-hours, Billion	Total energy used annually	UNDATA
Quality of life	Fraction	World database survey	UNDATA

Source Author's work

10.3.1 Framing of Hypothesis

The current research work makes an attempt to study the constructs that contribute towards renewable energy output. The literature review conducted in this paper helped deduce the following hypothesis:

H1: Economic growth (GDP) and foreign direct investment (FDI) share a significant positive relationship with renewable energy output.

Shahbaz et al. (2015) studied the association between foreign direct investment (FDI) and carbon footprints only to conclude that global environmental degradation has enhanced with increase in FDI.

Lau et al. (2014) analysed the link shared by growth in economy and carbon emissions to reveal that business, GDP, and FDI are crucial predictors of carbon emissions.

H2: Carbon tax, technological innovation, R&D expenses, and government expenses share a positive association with renewable energy output.

Mughal et al. (2022) examined the influence of innovation and R&D on enhancing energy efficiency by decreasing its intensity through a rigorous review of the environmental patents filed annually. The findings indicated that such patents help in improving energy efficiency and decrease its intensity. Lin and Li (2011) concluded that carbon/environment tax decreased energy usage and encouraged renewable energy consumption.

H3: The manufacturing sector expansion has a positive relationship with sustainable energy output.

Ouyang and Lin (2015) suggested that the Chinese manufacturing sector and its energy intensity are responsible for reducing carbon emissions whereas the ongoing manufacturing processes contribute towards an increase in the level of pollutants in the environment.

Wen et al. (2017) highlighted that the metals and minerals sector and its related activities led to increase in carbon emissions both in the short and long run.

H4: Renewable and non-renewable energy generation are negatively associated with each other but renewable energy generation and usage are positively linked to one another.

Goh (2014) established a negative link between non-fossil and fossil energy consumption. Cederborg and Snöbohm (2016) proved that total energy usage resulted in an increase in carbon emissions but usage of non-fossil fuels reduced carbon emissions.

H5: Quality of life and research publications are positively linked with sustainable power generation.

Levinson (2018) studied various social schemes and environmental measures and estimated that quality of life and renewable energy generation are positively associated with each other as non-fossil fuel consumption leads to reduction in pollution.

Beja (2012) also concluded that hot weather and air pollution massively negatively impacted the quality of life of people in Latin America.

10.3.2 Research Framework

The variables impacting sustainable energy generation in every included countries have been analysed by employing a panel data model to examine the fixed and random effects in this research work. Sample groups can maintain separate intercepts to explain each of the groups in a panel data. Adjustment quotient helps estimate the level of speed. The Ordinary Least Square (OLS) assists in the evaluation of the least variance among regression parameters like the error term, equal variance and uncorrelation. Thus, the equation representing the model is:

$$Y_{it} = \alpha_0 + \beta_1 X_{1,it} + \beta_2 X_{2,it} + \ldots + L + \beta N X_{n,it} + \varepsilon_{it} \tag{10.1}$$

i denotes the country, t is time, α_0 is constant, β is the regression coefficient, and ε_{it} is the error term.

The time series difference between the countries and the overall time and group effect can be employed to design the fixed effect model. The equation of the same is as follows:

$$Y_{it} = \alpha_0 \sum_{j=1}^{N} \alpha_1 d_{jt} + \sum_{r=1}^{T-1} \gamma_t e_{ri} \sum_{k=1}^{K} \beta_k X_{k,it} + \varepsilon_{jt} \tag{10.2}$$

α_0 is the fixed term intercept and α_1 represents intercept coefficient. $\sum_{j=1}^{N} \alpha_1 d_{jt}$ stands for country specific data and constants and γ_t is intercept of the year t. $\sum_{r=1}^{T-1}$ is the time constant.

Similarly, the random effect model can be designed based on time series difference and keeping the intercept stochastic. The equation for the same is the following:

$$Y_{it} = \alpha_1 + \sum_{K=1} K \ \beta_k X_{k,it} + \varepsilon_{it} = (\alpha + \mu_i + \gamma_t) + \sum_{K=1} K \ \beta_k X_{k,it} + \varepsilon_{it} \tag{10.3}$$

where γ_t shows the time difference.

Hausman test (Hausman 2019) was employed to select the fixed and random effect models. It suggests in case the intercept term is associated with the construct then fixed effect model is used and H_1 is accepted or else random effect model is used and H_0 is accepted. The equation for the same is:

$$H = (\beta_F - \beta_R)/[\mathrm{Var}(\beta_F) - \mathrm{Var}(\beta_R)] - 1(\beta_F - \beta_R) \sim X^2(k) \tag{10.4}$$

β_F is fixed effect model parameters, β_R is random model effect parameters, Var(β_R) stands for fixed effect model variance, Var(β_R) refers to random effect model variance, and $X^2(k)$ represents Chi-square value along with k degree of freedom.

The following equation explains the adjustment quotient used to estimate the level of speed:

$$Y_{it}^* = \alpha_1 . \, \mathrm{GDP}_{it} + \ldots + \alpha_{n.\mathrm{Nt}} + \varepsilon_{it} \tag{10.5}$$

Y^*_{it} stands for the adjustment quotient.

10.4 Empirical Results and Analysis

Table 10.3 represents standard deviations in case of responding variables are more when compared to the overall average which suggests high dispersion among variables. The standard deviations of GDP, carbon/environment tax, technology and innovation, government expenses, manufacturing growth, fossil fuel usage, and number of research articles published are more than the average. Hence, these variables are very discrete, while the rest are less discrete and more concentrated.

The solar and wind energy generation details of all the included countries are represented by Tables 10.4 and 10.5, respectively.

The solar and wind energy generation details regarding the dependent variable of all the included countries are represented by Tables 10.6 and 10.7, respectively.

The total solar and wind energy production is positively significantly influenced by GDP. Hence, H1 is accepted. Moreover, countries with a population of more than 10 million are found to impact wind energy production in a positive and significant manner. So do countries with rank greater than 20 and education index larger than 85%. Thus, it is in alignment with the hypothesis. Shahbaz et al. (2015) concluded that GDP or economic growth has an enormous impact on non-fossil fuel production. Growth in the economy results in increased generation of non-fossil fuels. Therefore, fossil fuel usage is reduced in due course of time (Kumar et al. 2024a, b).

The total solar and wind energy product is negatively influenced by carbon tax revenue. Hence, H2 is rejected. Environmental concerns are not solved by implementing policies related to the environment framed by the government. Carbon taxes and penalties did not suppress environmental degradation.

Technology and innovation has a significant positive influence on solar energy production (Kumar et al. 2024a, b) Hence, it is aligned with H2. Su and Moaniba (2017) indicated that R&D and innovation was highly related to climate change.

As per the results above, researched articles published annually share a significant positive relationship with solar and wind energy. It suggests that the number of research articles published in various journals assists in improving air quality index globally. Hence, H5 is accepted. Abeydeera et al. (2019) emphasised on research publications on sustainability from 2016 to 2020.

Table 10.3 Descriptive statistics

Constructs	Mean	Std.Dev.	Median	Min	Max
Solar energy output	2820.21	7551.28	83.33	0.00	51,263
Wind energy output	12,072.84	28,111.09	2610.23	0.00	2,31,500
GDP	14,124.97	30,381.24	3732.08	241.78	1,98,543.96
FDI	400.12	927.35	111.46	− 301.90	8412.56
Carbon/ environment tax	23,276.27	31,233.65	8248.89	0.00	1,35,412.27
Technology and innovation	4918.84	11,236.37	1279.09	4.29	59,870.07
R&D expenses	2.01	1.02	0.98	0.00	17.24
Health expenses	8.29	3.17	9.20	0.00	17.30
Government expenses	2607.16	4782.66	839.45	35.60	25,578.82
Manufacturing growth	3459.24	5891.71	988.38	0.00	34,014.00
Non fossil fuel usage	17.10	15.50	14.10	0.00	59.23
Usage of fossil fuel	4125.50	7790.34	897.78	22.06	44,893.24
Quality of life	6.39	3.21	7.22	0.00	8.61
Research articles published	40,263.60	79,309.72	13,124.40	311.10	5,20,892.39

Source Author's Work

Health and R&D expenses are negatively linked with the total solar and wind power production. Hence, it coincides with H2. Bawaneh et al. (2012) highlighted that money spent on the construction of hospitals globally has resulted in an increase in pollutant levels in the environment.

The growth of the manufacturing sector is negatively associated with renewable power generation. Hence, H3 is rejected. Bouckaert et al. (2021) showcased that for efficiency manufacturing units use thermal power and nor wind or solar power.

Renewable energy consumption and total renewable power production are positively linked with total solar and wind energy generation. Similarly, fossil fuel generation has a negative relationship with non-fossil fuel generation. Hence, H4 is accepted.

Adjustment factor value should be between 0 and 1 as per data analysis. Estimated target value can be timely achieved with a large value. As per Table 10.8, adjustment factor of solar energy generation is significantly consistent with the rest of the parameters of non-fossil power production for all the countries. But, in case of wind energy generation, all the adjustment coefficients are significantly consistent except for countries ranking above 20. Considering overall performance, wind energy

Table 10.4 Solar energy specifications

Test sample	H0:OLS		H0:one-way fixed		H0:OLS		H0:OLS		H0:one way random		H0:two way random	
	H1:one way fixed		H1:two way fixed		H1:one way random		H1:two way random		H1:one way fixed		H2:two way fixed	
	F	Sig	F	Sig	LM	Sig	LM	Sig	Hausman	Sig	Hausman	Sig
All countries	8.21	0.00	2.21	0.01	10.98	0.01	90.12	0.00	88.14	0.00	10.64	0.71
Pop. > 10 mn	8.45	0.00	3.63	0.00	7.11	0.09	87.37	0.00	35.80	0.00	8.19	0.82
Pop. < 10 mn	8.67	0.00	2.84	0.05	11.10	0.00	47.98	0.00	57.23	0.01	8.02	0.75
Country rank > 20	5.32	0.00	4.09	0.00	21.73	0.00	43.84	0.00	45.27	0.00	9.08	0.83
Country rank < 20	6.66	0.00	1.95	0.02	13.46	0.00	52.41	0.00	36.79	0.01	18.23	0.72
Edu Index > 85%	8.27	0.00	2.34	0.04	12.57	0.00	67.94	0.00	79.23	0.00	12.56	0.85
Edu Index < 85%	7.17	0.00	3.75	0.00	11.96	0.00	51.59	0.00	32.48	0.00	9.74	0.65

Source Author's Work

Table 10.5 Wind energy specifications

Test sample	H0:OLS		H0: one-way fixed		H0:OLS		H0:OLS		H0: one way random		H0: two way random	
	H1: one way fixed		H1: two way fixed		H1: one way random		H1: two way random		H1: one way fixed		H2: two way fixed	
	F	Sig	F	Sig	LM	Sig	LM	Sig	Hausman	Sig	Hausman	Sig
All countries	8.21	0.00	2.21	0.01	10.98	0.01	90.12	0.00	88.14	0.00	10.64	0.71
Pop. > 10 mn	8.45	0.00	3.63	0.00	7.11	0.09	87.37	0.00	35.80	0.00	8.19	0.82
Pop. < 10 mn	8.67	0.00	2.84	0.05	11.10	0.00	47.98	0.00	57.23	0.01	8.02	0.75
Country Rank > 20	5.32	0.00	4.09	0.00	21.73	0.00	43.84	0.00	45.27	0.00	9.08	0.83
Country Rank < 20	6.66	0.00	1.95	0.02	13.46	0.00	52.41	0.00	36.79	0.01	18.23	0.72
Edu Index > 85%	8.27	0.00	2.34	0.04	12.57	0.00	67.94	0.00	79.23	0.00	12.56	0.85
Edu Index < 85%	7.17	0.00	3.75	0.00	11.96	0.00	51.59	0.00	32.48	0.00	9.74	0.65

Source Author's Work

Table 10.6 Solar energy production index

Construct	All countries	Pop. > 10 mn	Pop. < 10 mn	Country rank > 20	Country rank < 20	Education Index > 85%	Education Index < 85%
	TWR	TWR	TWR	TWR	TWR	OWR	TWR
One adjustment factor	0.412(000)	0.343(000)	0.507(000)	0.472(000)	0.634(000)	0.345(000)	0.212(000)
GDP	0.450(000)	0.564(000)	– 0.078(000)	0.296(000)	0.189(000)	0.288(000)	– 0.376(000)
FDI net Inflow	0.245(000)	0.532(000)	0.134(000)	0.278(000)	– 0.431(000)	– 0.477(000)	0.147(000)
Carbon Tax-revenue	– 0.123(000)	– 0.135(000)	– 0.055(000)	– 0.175(000)	0.023(000)	– 0.103(000)	0.219(000)
Tech.& Innovation	0.610(000)	0.781(000)	– 0.112(000)	0.257(000)	0.009(000)	– 0.203(000)	0.562(000)
R&D expenses	3.124(000)	5.231(000)	7.890(000)	3.599(000)	8.248(000)	6.177(000)	10.343(000)
Health expenses	– 5.704(000)	– 4.332(000)	– 3.521(000)	– 5.553(000)	– 6.724(000)	– 7.434(000)	– 3.223(000)
Gov. expenses	– 1.001(000)	– 2.146(000)	3.256(000)	– 1.358(000)	– 4.612(000)	– 2.302(000)	– 1.146(000)
Manufacturing growth	– 0.895(000)	– 0.780(000)	0.673(000)	– 0.270(000)	– 0.347(000)	0.485(000)	0.391(000)
Non fossil fuel usage	10.240(000)	13.790(000)	16.683(000)	14.451(000)	12.268(000)	17.810(000)	18.903(000)
Usage of fossil fuel	– 2.009(000)	– 1.153(000)	– 3.304(000)	– 1.001(000)	2.144(000)	– 2.354(000)	– 4.121(000)
Quality of life	0.421(000)	0.373(000)	0.517(000)	0.492(000)	0.614(000)	0.325(000)	0.202(000)
Research articles published	0.440(000)	0.574(000)	– 0.088(000)	0.276(000)	0.159(000)	0.218(000)	– 0.336(000)

Source Author's Work

Table 10.7 Wind energy production index

Construct	All countries	Pop. > 10 mn	Pop. < 10 mn	Country Rank > 20	Country Rank < 20	Education Index > 85%	Education Index < 85%
	TWR	TWR	TWR	TWR	TWR	OWR	TWR
One adjustment factor	0.453(000)	0.356(000)	0.554(000)	0.465(000)	0.643(000)	0.337(000)	0.221(000)
GDP	0.640(000)	0.474(000)	− 0.128(000)	0.326(000)	0.198(000)	0.209(000)	− 0.367(000)
FDI net inflow	0.257(000)	0.523(000)	0.143(000)	0.287(000)	− 0.413(000)	− 0.468(000)	0.174(000)
Carbon Tax-revenue	− 0.134(000)	− 0.121(000)	− 0.044(000)	− 0.157(000)	0.032(000)	− 0.112(000)	0.291(000)
Tech. and innovation	0.639(000)	0.724(000)	− 0.121(000)	0.272(000)	0.016(000	− 0.232(000)	0.526(000)
R&D expenses	3.142(000)	5.213(000)	7.908(000)	3.059(000)	8.284(000)	6.169(000)	10.435(000)
Health expenses	− 5.740(000)	− 4.323(000)	− 3.512(000)	− 5.535(000)	− 6.742(000)	− 7.443(000)	− 3.232(000)
Gov. expenses	− 1.012(000)	− 2.164(000)	3.265(000)	− 1.385(000)	− 4.621(000)	− 2.320(000)	− 1.164(000)
Manufacturing growth	− 0.859(000)	− 0.708(000)	0.637(000)	− 0.207(000)	− 0.347(000)	0.458(000)	0.319(000)
Non fossil fuel usage	10.204(000)	13.719(000)	16.629(000)	14.415(000)	12.286(000)	17.801(000)	18.930(000)
Usage of fossil fuel	− 2.090(000)	− 1.135(000)	− 3.348(000)	− 1.016(000)	2.136(000)	− 2.345(000)	− 4.142(000)
Quality of life	0.412(000)	0.337(000)	0.525(000)	0.429(000)	0.641(000)	0.352(000)	0.224(000)
Research articles published	0.404(000)	0.548(000)	− 0.062(000)	0.267(000)	0.195(000)	0.281(000)	− 0.363(000)

Source Author's Work

production speed adjustment is faster when compared to solar energy generation in the non-fossil fuel sector.

10.4.1 Group Comparison

The authors have tried to explain the findings of explanatory constructs when compared to specifications of non-fossil fuel energy output. GDP with respect to wind energy generation has a significant negative influence on classifications with populations below 10 mn and nations below the ranking of 20. Hence, developing nations can make use of other forms of non-fossil fuels for power production in order to reduce the cost of wind energy generation.

In the classification of education index > 85%, FDI negatively influences solar and wind energy generation. Thus, developed nations can make use of FDI and avail the benefits of less expensive and penetrative nuclear energy generation to meet their economic and manufacturing requirements.

Carbon taxes have a significant and constant influence on solar energy generation. Countries with rank below 20 and education index below 85% have a positive influence on the same and thus, aligns with the hypothesis. In case of wind energy, the impact is overall significantly negative.

R&D expenses and innovation have a high positive influence on solar and wind power production. So, lack of research and technology can result in reduced use of solar and wind power.

Currently, health expenses have a significant negative influence on non-fossil energy generation. Healthcare centres are the largest source of carbon emission globally. Government expenditure and solar and wind energy output share a significant negative relationship with each other. With respect to population above 10 mn, education index below 85%, and country ranking above 20. Hence, in developed nations the reduction in government expenses leads to reduction in renewable energy mostly because, government tends to spend money on fossil fuel subsidies.

Manufacturing growth is positively associated with solar energy output for education index above 85% and thus, aligns with the hypothesis. But in case of wind energy production, the relationship is negative. Nuclear energy can be used as a substitute to fulfil industrial objectives in a country.

Wind and solar power production has a negative association with renewable power consumption for nations below 10 mn population. Thus, countries with fewer people consume more renewable power but it is limited to solar and wind power.

Published research articles and quality of life show results that are consistent with the hypothesis.

Table 10.8 Adjustment factor classification

Adjustment factor	All countries	Population > 10 mn	Population < 10 mn	Country ranking > 20	Country ranking < 20	Education Index > 85%	Education Index < 85%
Solar energy generation	0.592	0.700	0.453	0.529	0.469	0.708	0.874
Wind energy generation	0.879	0.890	0.778	− 0.674	0.614	0.913	0.825

Source Author's Work

10.5 Conclusion

On the basis of data collected from 32 countries within 2000–2023, the paper employed panel data to examine the fixed and random effect models and decode the various variables influencing sustainable energy production and economic growth of these nations.

This study makes an attempt to analyse the association shared by renewable energy generation and variables related to social and economic prosperity in a country. These factors have been categorised with respect to a few parameters, namely, countries with larger or less than 10 million population, countries ranking 20 and below, and countries with education index more or less than 85%. Adjustment coefficient analysis further investigates the effect of these constructs on renewable energy production in the included countries.

The findings proved that GDP, technological innovations, government expenditure, non-fossil fuel consumption, and published research articles have a positive and significant relationship with renewable energy production. Carbon tax, health expenditure, and manufacturing growth impact renewable energy production negatively. Hence, environmental related expenses will definitely affect non-fossil fuel usage. There are countries as well that opt for cheaper and more stable energy sources and use non-fossil fuel as a backup.

Speed level is estimated by the adjustment factor. As per the solar power generation adjustment coefficient, the classification of education index below 85% is the largest, and the population below 10 mn is the smallest. As per wind power generation adjustment coefficient, the classification of population above 10 mn is the largest, and the country with rank below 20 is the smallest. Thus, countries with a large population and higher rank can enhance renewable energy production. Moreover, countries with low population and rank require more time to enhance their renewable energy generation.

The issue of global warming and climate change has become a global roar in the twenty-first century. Ever since the 2000s, the concern has been observed to be accelerated by human production. Hence, reducing the pace of global warming and climate change is the major challenge faced by the human race on earth today. So, governments should take immediate measures and initiatives by framing a regulatory framework to mitigate exploitation of natural resources by people.

The study implicates that governments should plan carbon taxes sincerely and efficiently implement them on the grass root levels. Steps should also be undertaken by governments to enhance renewable energy production, renewable energy usage, and decrease carbon emissions. Thus, a fixed price and quantity system should be executed by authorities to improvise upon technology and decrease costs. This will ensure that non-fossil fuel energy can be a substitute for traditional energy in the future. Developed countries and international organisations should provide emerging economies with incentives to motivate them to buy the latest equipment and produce renewable power. Corporations can also do their bit to make their business environment friendly. These suggestions if incorporated in a realistic manner will help rectify errors in the market. Sustainability is a novel domain for academics and researchers and it demands further study and research. Therefore, this paper can act as a reference point for governments and administrations to highlight environmental problems and reduce their impact on society.

Further research is suggested on more varied variables. Studies on sustainability are not just restricted to solar, wind, and hydropower but they can also incorporate carbon dioxide, carbon monoxide, methane, and other greenhouse gases to enable the idea of a green environment.

Data Availability Statement NA.

Conflict of Interest There is no conflict of interest.

References

Apergis N, Payne JE (2010) Renewable energy consumption and economic growth: evidence from a panel of OECD countries. Energy Policy 38(1):656–660

Bawaneh A, Massadeh D, Akour I, Haija AA, Alshurideh M (2012) The impact of green auditing on organizational performance in Jordan: the moderating effect of the auditor's opinion

Beckerman W (1992) Economic development and the environment: conflict or complementarity? World Development Report, 1–50

Beja EL (2012) Subjective well-being approach to environmental valuation: evidence for greenhouse gas emissions. Soc Indic Res 109(2):243–266

Bernauer T, Koubi V (2011) States as providers of public goods: how does government size affect environmental quality? SSRN Electron J

Brunnermeier SB, Cohen MA (2003) Determinants of environmental innovation in US manufacturing industries. J Environ Econ Manag 45(2):278–293

Cederborg J, Snöbohm S (2016) Is there a relationship between economic growth and carbon dioxide emissions?

Charles Rajesh Kumar J, Majid MA (2020) Renewable energy for sustainable development in India: current status, future prospects, challenges, employment, and investment opportunities. Energy Sustain Soc. BioMed Central Ltd.

Daggash H, Dowell N (2019) Higher carbon prices on emissions alone will not deliver the Paris agreement. Joule.

Damette O, Marques AC (2019) Renewable energy drivers: a panel cointegration approach. Appl Econ 51(26):2793–2806

Diener E (2013) The remarkable changes in the science of subjective well-being. Perspect Psychol Sci 8(6):663–666

Goh E (2014) The modes of China's influence: cases from Southeast Asia. Asian Surv 54(5):825–848

Goh T, Ang BW (2020) Four reasons why there is so much confusion about energy efficiency. Energy Policy 146

Grossman and Krueger (1995) Economic growth and the environment. Quart J Econ 110:353–377

Hausman C (2019) Market deregulation and nuclear safety. Energy Econ 82:68–69

Hsin-Ning S, Moaniba IM (2017) Does innovation respond to climate change? Empirical Evidence from Patents and Greenhouse Gas Emissions, Technological Forecasting and Social Change 122:49–62

Hu J, Tang Q, Wu Z, Zhang B, He C, Chen Q (2023) Optimization and assessment method for total energy system retrofit in the petrochemical industry considering clean energy substitution for fossil fuel. Energy Convers Manage 284:116967

Jeong JS, Kim YS, Redillas MC, Jang G, Jung H, Bang SW, Choi YD, Ha SH, Reuzeau C, Kim JK (2013) OsNAC5 overexpression enlarges root diameter in rice plants leading to enhanced drought tolerance and increased grain yield in the field. Plant Biotechnol J 11(1):101–114

Jeong JI, Park RJ, Song CK, Yeh SW, Woo JH (2024) Quantitative analysis of winter PM2.5 reduction in South Korea, 2019/20 to 2021/22: contributions of meteorology and emissions. Sci Tot Environ 907

Kumar R, Khanna R (2023) RPA (Robotic Process Automation) in finance & accounting and future scope. In: 2023 second international conference on augmented intelligence and sustainable systems (ICAISS). IEEE, pp 1640–1645

Kumar R, Kandpal B, Ahmad V (2023a) Industrial IoT (IIOT): security threats and countermeasures. In: 2023 international conference on innovative data communication technologies and application (ICIDCA). IEEE, pp 829–833

Kumar R, Malholtra RK, Grover CN (2023b) Review on artificial intelligence role in implementation of Goods and Services Tax (GST) and future scope. In: 2023 international conference on artificial intelligence and smart communication (AISC). IEEE, pp 348–351

Kumar R, Sexena A, Gehlot A (2023c) Artificial intelligence in smart education and futuristic challenges. In: 2023 international conference on disruptive technologies (ICDT). IEEE, pp 432–435

Kumar R, Lande A, Kumar D, Malhotra RK, Sharma A (2023d) Technology bridging-in entrepreneurs and consumers in product development. In: 2023 11th international conference on intelligent systems and embedded design (ISED). IEEE, pp 1–5

Kumar R, Saxena A, Singh R (2023e) Robotic process automation bridge-in banking institute and consumers. In: 2023 international conference on disruptive technologies (ICDT). IEEE, pp 428–431

Kumar V, Kumar R, Goel R, Singh T, Kaur MJ (2024a) The evolution of digital marketing technologies: a fuzzy TOPSIS approach to selected technology in E-commerce platform. Int J Syst Assur Eng Manag 1–14

Kumar R, Goel R, Singh T, Mohanty SM, Gupta D, Alkhayyat A, Khanna R (2024b) Sustainable finance factors in Indian economy: analysis on policy of climate change and energy sector. Fluctuation Noise Lett 23(2)

Lan J, Kakinaka M, Huang X (2012) Foreign direct investment, human capital and environmental pollution in China. Environ Resour Econ 51:255–275

Lau LS, Choong CK, Eng YK (2014) Investigation of the environmental Kuznets curve for carbon emissions in Malaysia: do foreign direct investment and trade matter? Energy Policy 68:490–497

Leuschner P (2014) The effect of GDP per capita on renewable energy production in China. University of CO, Boulder, Undergraduate Honors Theses

Levinson A (2018) Mercury and air toxics standards: co-benefits and the courts in US cost-benefit analysis, case studies in the environment

Lin B, Li X (2011) The effect of carbon tax on per capita CO_2 emissions. Energy Policy 39(9):5137–5146

Lopes AR (2014) The impact of social media on social movements: the new opportunity and mobilizing structure. J Polit Sci Res 2–23

Lopes M, Serrano L, Ribeiro I, Cascão P, Pires N, Rafael S et al (2013) Emissions characterization from EURO 5 diesel/biodiesel passenger car operating under the new European driving cycle. Atmos Environ 84:339–348

Lopez RE, Palacios A (2010) Have government spending and energy tax policies contributed to make Europe environmentally cleaner?

Mert M, Bölük G, Çağlar AE (2019) Interrelationships among foreign direct investments, renewable energy, and CO_2 emissions for different European country groups: a panel ARDL approach. Environ Sci Pollut Res 26:21495–21510

Mohanty SN, Singh T, Goel R, Baral SK, Kumar R (2024) A study on building awareness in cyber security for educational system in India using interpretive structural modellings. Int J Syst Assur Eng Manag 1–11

Mughal N, Arif A, Jain V, Chupradit S, Shabbir MS, Ramos-Meza CS, Zhanbayev R (2022) The role of technological innovation in environmental pollution, energy consumption and sustainable economic growth: evidence from South Asian economies. Energy Strategy Rev 39

Ouyang X, Lin B (2015) An analysis of the driving forces of energy-related carbon dioxide emissions in China's industrial sector. Renew Sustain Energy Rev 45:838–849

Bouckaert S, Pales AF, McGlade C, Remme U, Wanner B, Varro L, D'Ambrosio D, Spencer T (2021) Net zero by 2050: a roadmap for the global energy sector

Raheem ID, Ogebe JO (2017) CO2 emissions, urbanization and industrialization: evidence from a direct and indirect heterogeneous panel analysis. Manag Environ Qual Int J 28(6):851–867

Sarkodie SA, Strezov V (2019) Effect of foreign direct investments, economic development and energy consumption on greenhouse gas emissions in developing countries. Sci Total Environ 646:862–871

Shahbaz M, Nasreen S, Abbas F, Anis O (2015) Does foreign direct investment impede environmental quality in high-, middle-, and low-income countries? Energy Econ 51:275–287

Sinha A, Shahbaz M (2018) Estimation of environmental Kuznets Curve for CO_2 emission: role of renewable energy generation in India. Renew Energy 119:703–711

Stern DI (2004) Economic growth and energy. Encycl Energy 2(00147):35–51

Stern DI (2004) The rise and fall of the environmental Kuznets curve. World Dev 32(8):1419–1439

Udara Willhelm Abeydeera LH, Wadu Mesthrige J, Samarasinghalage TI (2019) Global research on carbon emissions: a scientometric review. Sustainability 11(14):3972

Wen Z, Song K, Shang Y, Fang C, Li L, Lv L, Chen L (2017) Carbon dioxide emissions from lakes and reservoirs of China: a regional estimate based on the calculated pCO_2. Atmos Environ 170:71–81

Withana S, Ten Brink P, Kretschmer B, Mazza L, Hjerp P, Sauter R (2013) Evaluation of environmental tax reforms: international experiences

Zhongping W, Changliang S, Qiang L, Gang W (2011) Coupling trending analysis about urbanization and urban resource in Beijing. Energy Procedia 5:1589–1596

Chapter 11
Green Bonds and Climate Bonds Policies at the International Level

Pooja Devi

Abstract In the world of finance, due to global environmental challenges, there is big transformative financial tools such as climate bonds and green bonds have come in existence. In the growing trends of sustainable finance, these climate and green bonds are being utilized for reducing carbon emissions and increasing energy efficiency to promote sustainable development. This chapter includes History and rise of greens and climate bonds. This study also explores the Innovations and mechanism of both kind of bonds. International standards such as European green bond standard, ASEAN green bond standard, green bond principles and ISO 14030 for green and climate bond also being taken in this study. This chapter explore the role of green and climate bond in examining the international policies and framework for the purpose of low carbon economy. In this chapter we also include the roadmap of bonds which explore the whole mechanism of green and climate bond from project identification to monitoring and compliance toward it. This study analysis the role of international organizations such as ICMA, MDBs. These organization formulate framework for making principles and policies for green bond and climate bond. The aim of this study is to provide comprehensive understanding of role of green and climate bond in sustainable development at global level and to give valuable insights for organizations and policy makers toward sustainable future.

Keywords Green bond · Climate bond · International organizations · Policies · Environment

P. Devi (✉)
Uttaranchal Institute of Management, Uttaranchal University, Dehradun, India
e-mail: Poojaoneworld137@gmail.com

R. Kumar (ed.), *Sustainable Finance*, Sustainability Solutions,
https://doi.org/10.1007/978-3-031-97962-0_11

11.1 Introduction

In the recent years, there has been urgent attention toward the tackle climate change which has become the center of global financial and policy debates. The new financial tools like green bonds and climate bonds, designed to raise funds for eco-friendly projects, which help in reducing the carbon emissions and lessen environment damage. The financial system can achieve environmental goals through these innovative bonds. The green bond is a type of loan that is specially designed for money- raising meant to fund environment-friendly projects. The growth in demand for green bonds reflects a broader trend in the investment community. Pension funds and asset managers are increasingly institutional investors that seek out green bonds as part of their ESG strategies (Kumar and Khanna 2023). Green bonds represent a possible pathway for portfolios to go from being carbonized to being aligned to sustainability goals, coupled with, of course, financial returns. Investors also appreciate the fact that green bonds are more financial competitive than regular bonds (Flaherty et al. 2017). Most of them are issued by very creditworthy entities, which offer both environmental impact and security to investors. In addition, green bonds ensure portfolio diversification, thereby removing risks of exposure from climate-related risk in a low-carbon transition economy. Climate bonds are the type of green bond, which emphasizes on projects that combat climate change, such as proper building of transportation, renewable energy, enhancement of the efficiency of energy. The popularity has been on the rise because governments, corporations, and investors see their potential to support climate efforts and generate a profit (Bhutta et al. 2022).

11.1.1 History of Green and Climate Bonds

From a niche financial product, to now achieving mainstream status as an investment tool for sustainability, the story of green bonds has borne testimony to the exponentially increasing efforts around the globe to combat climate change. The first green bond came in 2007 and the concept of green finance began to mature. Development banks were the early adopters, with governments soon following up, raising funds through green bonds for environmental projects. Green bond principles introduced in 2014 for standardized the market and boosting investor confidence (Kumar et al. 2023a, b, c, d, e).

11.1.1.1 Timeline and Growth Trends

The 2015 Paris Agreement formed a necessary step toward great milestones that talked about the infusion of private capital toward reaching the goals for climate. By 2016, green bonds had actually taken off and gained lots of momentum, with both corporate and municipal issuers jumping into the fray. Now, the market has also extended to include projects in renewable energy and efficiency, clean transportation, and sustainable agriculture (Ballester et al. 2024). Climate bonds, which are targeted at projects specifically focused on climate mitigation and adaptation, also feature strongly in the market. The growth is consistent with urgent needs for targeted financing in response to specific kinds of climate challenge (Kumar et al. 2023a, b, c, d, e).

11.1.2 The Rise of Green Bonds and Climate Bonds

As we all know, First of all European investment bank issued green bonds for setting the stage for a global green finance market. By increasing awareness of climate risks as well as growing recognition among the financial sector that it has its role to play in making this sustainability a reality, green bond issuance had crossed $500 billion by 2021. The Paris agreement of 2015 also marked a turning point by focusing on the need for capital investment on an unprecedented scale (Cheong and Choi 2020). It breathed green finance, especially bond financing, into the life of preferred capital-raising for environmental projects. Green bond issuances were encouraged through related regulatory frameworks established by governments and international organizations for the rapid growth of these instruments.

11.1.3 Innovation in Green Bond Markets

The growth of the green bond market is given by Innovation. New financial instruments, technology, and collaboration are transforming how green bonds are issued and traded. Apart from the traditional green bonds, newer instruments such as sustainability linked bonds and social bonds have emerged. These bonds enter this sphere of sustainable finance, combining financial performance with sustainability targets. Mechanism such as green securitizations and blended finance approaches add to the market's ability to mobilize funds for a variety of sustainability projects (Cortellini and Panetta 2021). As we all know, it increases transparency through blockchain technology, especially in the issuance and reporting on bond papers. Advanced data analytics is applied in measuring and reporting the environmental impacts of the projects with more accurate levels. Additionally, digital platforms ensure smoother processes in green bonds issuance and trading, hence making the market more accessible for issuers and investors. On other hand Collaboration among market participants is key to overcoming challenges in the green bond market. Knowledge sharing, best practices and innovative solutions are thus crucial functions of forums, industry conferences, and international initiatives driving market innovation.

11.2 Green and Climate Bonds: A Roadmap Overview

Green and climate bonds are crucial financial products through which funds are being channeled into environment-friendly sectors. They become essential resources for mitigating climate change, sustainability and building international cooperation in green finance (Otek Ntsama et al. 2021). This abridged version covers the major mechanisms, frameworks of regulations, and the principal international actions that have shaped the green bond market.

11.2.1 Mechanism of Green and Climate Bonds

- **Identification of Projects**: Green bond funds projects targeted toward particular environmental objectives. These projects reduce greenhouse gas emissions, enhance energy efficiency, and build climate resilience (Baker et al. 2022).
- **Green Bond Framework Development**: Issuers establish a Green Bond Framework that describes how the proceeds are to be allocated, project eligibility, and reporting. It also needs to align with the international guidelines, for example, the Green Bond Principles (GBP) or Climate Bonds Standard, so there is some level of accountability and transparency involved.
- **External Review and Certification**: For the most part, issuers seek third-party verification to ensure the environmental integrity of their bonds: such as second-party opinions, certifications (for example, from Climate Bonds Initiative), third-party audits for eliminating greenwashing issues, and also from an investor's end, trust building (Cioli et al. 2021).
- **Bond Issue**: Governments, corporate organizations, municipalities, or financial institutions are the issued parties of Green bonds often used by institutional clients such as pension funds, buy these bonds, which help channel funds into green projects.
- **Allocation of Funds**: Earnings will be accounted for and used in green projects only after the bonds have been issued. Funds are ring-fenced, which means they can only be used for the given projects, ensuring increased transparency and garnering much-needed investor confidence (Rumpf 2019).
- **Implementation Projects**: Green funds for these bonds are executed based on a structured framework, toward objectives such as adding new renewable energy capacity, better water resource management, decreasing pollution, or making the energy system more efficient.
- **Implementation Projects**: Green funds for these bonds are executed based on a structured framework, toward objectives such as adding new renewable energy capacity, better water resource management, decreasing pollution, or making the energy system more efficient (Logutov 2023).
- **Impact Measurement and Reporting**: Issuers provide quarterly or annual reports on how the funds have been allocated and resulting environmental

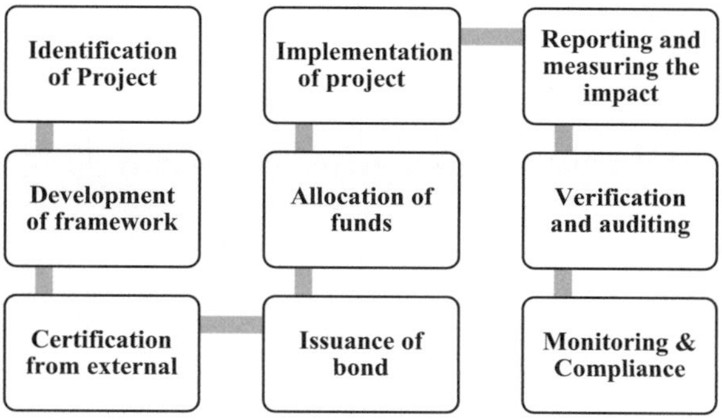

Fig. 11.1 Exploring the mechanism of green and climate bonds

impacts, including tons of CO_2 prevented, megawatt-hours of electricity saved, or megawatts of renewable energy generated.

- **Verification and Audits**: There's independent verification of fund allocation and impact. Report results are successively subject to third-party audits. The outcome and hence its credibility are assured by third-party risk. Third-party risk is therefore minimized by greenwashing.
- **Ongoing Monitoring and Compliance**: Issuers are also responsible for monitoring the progress of green projects to ensure that environment objectives are followed through throughout the life cycle of the bond. Periodic reporting helps keep the market aligned with its original set goals and enhances credibility in the market (Dell'Atti et al. 2022) (Fig. 11.1).

11.3 The Role of International Standards

The standardized standards set for green and climate bonds have played a key role in their credibility. In 2014 International Capital Market Association (ICMA) was the initiators of green bond principles to provide voluntary guidelines for the issue of green bond. The principles ensure that the application proceeds of the green bonds are completely used only for environmental projects, there is also measurement and reporting of environmental benefits. The Climate bonds initiative (CBI) established a certification scheme for climate bonds, outlining rigorous standards for qualifying projects in its taxonomy (Fatica et al. 2021). This will prevent "greenwashing", which is the labeling of non-sustainable projects as green in order to attract investments. Trust among issuers will be one result of these standards, and investors can be confident that their investments are directed toward meaningful environmental projects (Kumar et al. 2024a, b).

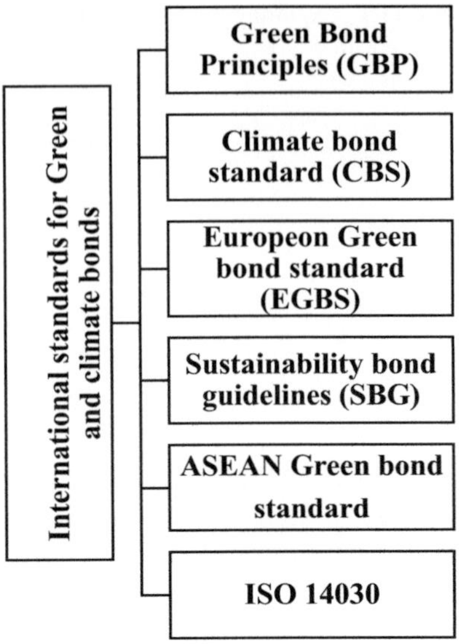

Fig. 11.2 Exploring the international standards for green and climate bonds

11.3.1 Global Policy Initiatives and Cooperation

Global cooperation is essential to the success of green and climate bonds. Recent initiatives have seen Multilateral development banks (MDBs) of the World Bank others around the world, like the European bank for Reconstruction and Development (EBRD) and Asian Development Bank (ADB), take the lead in issuing green bonds, financing large-scale environment projects around the globe. Other policy frameworks that will take center stage are the SFDR by the European Union and the Taxonomy Regulation (Flammer 2020). All these will drive transparency in investments and accountability. Green Deal, an initiative by the EU to be carbon neutral by 2050, is another driver for green bond issuances. Recent international efforts, such as the Network for Greening the Financial System (NGFS) and the United Nations Environment Program Finance Initiative (UNEP FI), have aligned policies that make green bonds tradable across borders (Fig. 11.2).

11.3.2 Role of International Organizations

International organizations also play an integral part in green and climate bond growth. The Climate Bonds Initiative (CBI) has made certifications of climate bonds more stringent so that the proceeds are spent efficiently toward the cause of controlling climate change. International organizations, such as the United Nations, have further promoted global best practice in sustainable finance through initiatives like Principles for Responsible Investments (PRI) and UNEP FI (Krueger et al. 2020).

The above institutes, along with providing capital, also act as a guiding light for other issuers to emulate (Lebelle et al. 2020a, b).

11.3.3 Regulatory Frameworks and Policies

Maturing markets for green bonds require sound regulatory frameworks and policies in addition to credibility. International standards and regional policies ensure the integrity of such a market; their ongoing process of shaping regulatory challenges, such as standardization and transparency, continues to shape it (Kumar et al. 2024a, b).

11.3.4 Policy Challenges and Barriers

- **Standardization and Consistency**: There is a lack of uniformity in terms of definitions and standards across regions, hence complicating the market. International frameworks will come out straight if expanded to heighten expansion in the market.
- **Greenwashing**: All this despite guidelines set remains susceptible to greenwashing. All these call for verification mechanisms strong enough to hold onto integrity in the market.
- **Reporting and Transparency**: Reporting with transparency and accuracy forms a challenge. It will help boost investor confidence and, in turn, have the green bonds gain credibility.
- **Accessibility and Inclusion**: Accessibility and inclusivity are primarily important for the rise in participation of various issuers that participate in green bonds, such as small and medium-sized enterprises and developing nations [Tang et al.].
- **International Initiatives**: International organizations play critical roles in shaping green bond markets and driving global collaboration toward sustainable finance.

11.3.5 Multilateral Development Banks (MDBs)

The MDBs, which include the World Bank, the European Bank for Reconstruction and Development (EBRD), and the Asian Development Bank (ADB), have been at the forefront of green bonds.

- **Market Development**: MDBs were among the earliest adopters of green bonds. The world's first green bond was launched by the World Bank in 2008. Their leadership has set precedents for the market and improved investor confidence.

- **Standard-Setting**: MDBs contribute to the development and promotion of standards, such as the GBP and Climate Bonds Standard, and ensure that bonds meet strict environmental criteria.
- **Project Financing**: MDBs resort to green bonds to fund projects in renewable energy, energy efficiency, sustainable infrastructure, as well as climate resilience. This is to accelerate the transition to a low-carbon economy.
- **International Cooperation**: MDBs collectively work across borders on green finance through initiatives like Network for Greening the Financial System (NGFS) to enable financial systems to support climate goals.

11.3.6 United Nations and Climate Finance

The United Nations has motivated climate finance through different initiatives, such as developing sustainable investments and creating a green bond market.

- **UNEP Finance Initiative (UNEP FI)**: UNEP FI partners with financial institutions to drive sustainability into their operations. Their efforts and other efforts that feature the PRB converge into promoting green bonds.
- **UNFCCC and the Green Climate Fund**: UNFCCC created the GCF, which complements funding inputs in climate projects in developing countries, thus complementing the goal of climate finance, globally (Wang et al. 2022).

11.3.7 Sustainable Development Goals (SDGs)

Green and climate bonds facilitate the transition in meeting the UN's Sustainable Development Goals, particularly those that fall under climate action, clean energy, and sustainable infrastructure. In short Green and climate bonds are a powerful tool in financing climate change adaptation and sustainability projects. The mechanism of green and climate bonds ensures their transparency, accountability, and measurable environmental effects. Through credible regulatory frameworks and international initiatives, credit enhancement and growth for the market are maintained. The current challenges are standardization, greenwashing, and access-all on the table to ensure that the potential of green bonds is fully realized. As global cooperation on sustainable finance deepens, green bonds will remain a key tool for driving the low-carbon transition toward a more resilient future.

11.4 Green Bonds and Climate Bonds Policies at the International Level

11.4.1 Paris Agreement and Its Influence on Bond Policies

The Paris Agreement, introduced in 2015, is an international landmark accord focused on the limitation of global temperature rise. It has promoted the expansion of green and climate bonds, affecting climate finance, investment strategies, transparency, and the growth of sustainable finance markets (Yeow and Ng 2021).

- **Climate Finance Commitments**: Climate finance is among the most prominent components of the Paris Agreement, mainly expressed in Article 9 where developed countries agree to mobilizing $100 billion every year toward transitioning developing nations concerning efforts toward combating climate change (Gianfrate and Peri 2019a, b). This interest has already tried to put a tool for funding under green bonds toward green projects, and go-and-government and financial institutions now must adjust policies on emitting green bond.
- **Nationally Determined Contributions (NDCs)**: Countries, in the Paris Agreement, commit to presenting Nationally Determined Contributions, which are detailed climate action plans and targets. NDCs have lately been incorporated into investment policy and green bonds have become routine practice. Green bonds by embracing NDC objectives therefore allow countries to finance the realization of national and international climate goals through projects such as renewable energy inputs (Broadstock and Cheng 2019).
- **Enhanced Transparency and Disclosure**: One of the most significant features of the Paris Agreement is climate finance transparency and accountability. The agreement would, thus, entail monitoring and reporting of flows of climate finance, and such expenditures from green bonds spent how and what. For this reason, issuers of such bonds have to upgrade investor confidence by reporting on how the proceeds are invested and revealing environmental impacts of funded projects.
- **Market Growth and Innovation**: This has thus incubated innovation and growth in the green bond market as ambitious climate targets bring on growing demand for sustainable finance solutions. Filling the demand gaps are types of sustainable finance bonds such as ever more well-known sustainability bonds, social bonds, among others, such as more traditional Green Bonds- all confirming the broader climate objectives with which other regions are in alignment, such as that of the EU's Green Bond Standard (Baker et al. 2018).

11.5 Green Bonds in the Capital Market

Actually, green bond proceeds would mobilize revenues to be utilized for financing environmentally friendly projects. It raises so many questions on its role within the financial ecosystem.

- **Diversification of Investment Opportunities**: This will give investors a kind of asset that meets environmental and sustainability criteria. In this context, investment from pension funds and insurance companies in green bonds will not only diversify their portfolios. This approach promotes a portfolio performance together with the spread of risks in diversified sectors (Stellner et al. 2015).
- **Be a contributor to achieving SDG**: Green bonds were to issue finance on sustainable viable projects that will realistically and feasibly promote the United Nations Sustainable Development Goals, especially in climate action and clean energy. They can also be used by investors seeking evidence or proof of their sustainability commitment and to attract environment, social, and governance investors (Zerbib 2019). In truth, green bonds can fit perfectly under the SDGs to bridge private capital with global sustainability targets.
- **Market Liquidity and Growth**: This demand from investors and issuers has, in turn, flowed into liquidity for the green bonds and increased access to capital. As the market grows with time, it will improve liquidity for the green bonds to make pricing more efficient and doing business easier. Growth in green bonds reflects the trend toward sustainable finance at large, encompassing new bond structures as well as growth in sector-specific bonds (Bachelet et al. 2019).
- **Effects on Conventional Bonds Market**: Green bonds act as standards for transparency and environmental reporting. And so, they influence the conventional bond markets. Many issuers of traditional bonds consider incorporating sustainability considerations while building their offerings, which further contributes to an increase in other sustainable financial instruments in the space, such as social and sustainability-linked bonds.

11.5.1 Investor Demand for Green Financial Instruments

After the appearance of several key factors, demand for green financial instruments had risen significantly.

- **Raise awareness about the climate risk**: It raises them more aware and sensitive to the potential risks changes in climate might pose on their investments today-increased regulatory change and extreme weather events. Green bonds provide a fine opportunity to tap projects that mitigate such risks and contribute to the low-carbon economy (De Spiegeleer and Schoutens 2019). Hence, green bonds complement responsible investment goals well and benefit from the growing importance of ESG criteria in investment decisions.
- **Regulatory and Policy Framework**: Countries are adding more policies globally to increase sustainable finance and invest in green bonds. In this regard, the Sustainable Finance Disclosure Regulation of the European Union has provided greater inclusivity for investors so that they might make a more holistic review of sustainable investment. Climate-related elements in financial regulatory measures have also been incorporated by NGFS and the likes, paving further regarding green bond markets.

- **Demand from Institutional Investors**: The leading institutional buyers of green bonds are pension funds and asset managers. Many of them, increasingly, put their investments into ESG considerations and thus find financial instruments supporting the sustainable goals they pursue. Green bonds are a finance instrument with both monetization outcomes and an added positive environmental impact.
- **Performance and Return Expectations**: Their yields are now competitive with other conventional bonds, and their positive performance plus benefits for the environment improve their attractiveness as an investment tool that could balance financial goals with sustainability. Further, the maturation of the market should continue to make green bonds an increasingly attractive investment opportunity (Lebelle et al. 2020a, b).
- **Risk Management and Market Integration**: Effecting High-Quality Green Bonds in international financial markets: There must be risk management relating to associated risks, and the establishment of sound market integration.
- **Environmental Impact Assessment**: One of the biggest problems is that projects financed by green bonds produce the environmentally positive attributes intended. All these approaches encompass, above all, the Green Bond Principles-GBP-and the Climate Bonds Standards, which provide guidelines for independent verification of such impacts and, in general, keeping things transparent is all about investor confidence (Glomsrød and Wei 2018).
- **Financial Risk Management**: Risks of green bonds are like any other financial instrument, the concept of green bonds carries interest rate, credit and liquidity risks. Every investor will quantify those risks and, going forward, take into account various factors such as the credit worthiness of the issuers and the performance of projects that they finance (Tang and Zhang 2020). Management of those risks will sustain the stability and growth of the Green Bond market.
- **Market Integration and Standardization**: The markets have increased their comparability of green bonds through standardization about the inclusion of green bonds in the international markets (Gianfrate and Peri 2019a, b). Wider universal definitions, as well as verification and standard reporting requirements, assure comparability of green bonds while their inclusion in markets remains very attractive to investors, which spurs wider market participation through them.
- **Market development and Entrepreneurship promotion**: Continued growth in the green bond market will depend on further innovation and expansion, enabling more diverse sets of issuers and projects to become eligible for green finance. It supports new financial products, such as sustainability-linked bonds, that complement green bonds by also seeking to address wider environmental and social objectives.

11.6 Roles in Green Bonds in Mitigating Climate Change

- **Financing Renewable Energy**: These green bonds are playing a very important role in financing renewable energy sources, including solar, wind, hydro, and geothermal energy. The entire end reduces dependence on fossil fuel and decreases greenhouse gas emissions. Such a big project financed by a green bond, for instance, one aimed at solar farming, reduces carbon emissions because it replaces electricity by coal or natural gas-generated electricity.
- **Energy Efficiency Support**: Green bonds also finance energy efficiency projects, for example, retrofitting buildings with energy-saving technologies and upgrading industrial processes. While investments in energy efficiency reduce energy consumption, it typically strikes the environment less hard. Improved efficiency reduces emissions, thereby lowering the cost of energy consumption, inspiring resource sustainability.
- **Building climate resilience**: The bonds themselves-the green bonds-are adaptation projects to attain a form of resilience against climate change. These include funding for flood defenses, drought-resistant infrastructure, climate-resilient agriculture to safeguard vulnerable communities from climate-related disasters and protect livelihoods and infrastructure (Fig. 11.3).

11.6.1 Social Implications and Benefits

While green and climate bonds primarily focus on environmental outputs, they bring about enormous benefits that are socially synonymous.

- **Employment generation and economic development**: The funds created under renewable energy and energy efficiency enhancement initiatives through green bonds create employment opportunities in construction, manufacturing, and technological development sectors (Maltais and Nykvist 2020). Green bonds also stimulate economic growth hence boost regional economies, regionally so, in terms of the economy's stability.
- **Public Health and Well-being**: Improved public health is normally carried out by the projects initiated through green bonds. For instance, renewable energy

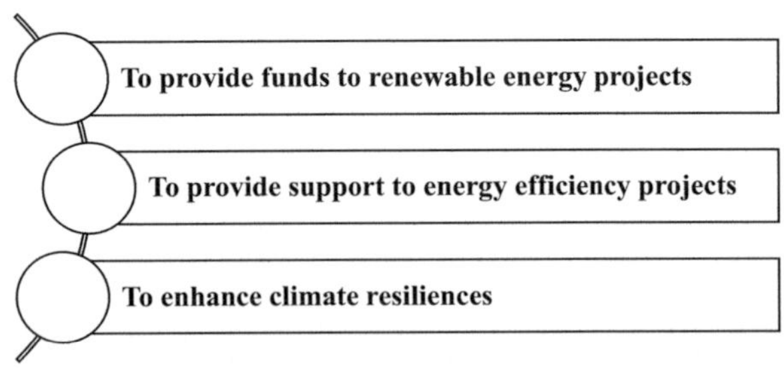

Fig. 11.3 Integrating the roles of green and climate bonds in mitigating climate change

improves good-quality air by minimizing the emissions linked with fossil fuel combustion. The same energy-efficient building tends to improve the internal environment by minimizing levels of pollutants and comfort.

- **Social Equity and Inclusion**: Green bonds open social equity in access to funds toward projects related to milestones in environmental justice. Offering clean energy to poorer communities or enhancing climate resilience in vulnerable territories will thereby decrease environmental inequities. Inclusive development advances by considering these disadvantaged groups.

11.7 Challenges and Future Directions

The green bond market is still, however, with many challenges it needs to overcome toward maximizing impact while growing.

- **Policy Gaps and Co-ordination Issues**: Poor harmonization of policies and regulatory frameworks remains the greatest challenge. Such fragmentation makes the issuance processes complicated and therefore inefficient to the issuers and the investors.
- **Patchy Regulations**: The requirements in the case of green bonds are differentiated by countries, making cross-border investments and compliance rather challenging where other regions have weaker guidelines while the EU has more intensive regulations.
- **Coordination Among Stakeholders**: Further growth depends upon coordination and cooperation among governments, regulatory agencies, and other financial markets that are involved. More inconsistencies in regulations arise from the divergent sustainability agendas and policies of countries.
- **Policy Gaps**: There is under-incentivized area that does not effectively support the issuing of green bonds. Such gaps must be counteracted by holistic policy action, for instance subsidies and incentives on green finance initiatives.

11.8 Scaling up Green Finance

To achieve its true potential, the market needs to scale up the green bonds market in order to mobilize more capital for climate-related projects.

- **Scaling green market participants**: The green market is quite concentrated with large corporations and governments. Scaling for the small and medium municipalities and private companies are massive ways to scale the green finance. Improving market infrastructure and processes that lead to issuance are critical ingredients to lure new participants.

- **Mobilizing Capital to Emerging Markets**: Emerging markets face severe barriers in accessing green finance. Barriers include, among others, market infrastructure and increased perceived risks for lending or investing in these countries. International organizations and MDBs can assist such regions by providing technical assistance, capacity building, and mitigating risks.
- **Rising Market Transparency**: Transparency is what will guarantee true investor confidence, and it forms the bedrock of which green bonds will indeed deliver on the environmental promise. It is because of this that enhanced reporting standards, clearer information on impacts, and rigorous verification processes would help address all these concerns. Increases in standard reporting frameworks help increase transparency together with third-party independent verification. Greater access toward data on impact also enhances third-party independent verification (Park 2018). All these steps can further ensure that the contributions of green bonds are on their way toward environmental and climate goals while pertinent information reaches investors for judicious investment decisions.

11.9 Case Studies, Published in December 2024

Several international case studies point to the increased involvement of green and climate bonds in helping to mitigate environmental issues.

- **China Green Bond Market**: It has been a world leader in green bond issuance so far. China is moving into 2024 funding renewable energy, pollution prevention, and energy efficiency with a robust framework for green bonds. Benefits have been accruing due to efforts put forward by financial institutions and regulators for the growing market of green finance in China.
- **European Union (EU) Development Programs**: The European Union stays at the top for climate bond issuance financing and all other diverse projects under its European Green Deal. Increased transparency for EU Taxonomy on Sustainable Activities and the European Green Bond Standard attracts this diversified investor, including pension funds and asset managers.
- **Brazil and Latin America**: Brazil's green bond market, financing sustainable agriculture and reforestation projects, is just taking off in 2024. Bonds will fund the decline in further Amazonian deforestation and greenhouse gas emissions. As Brazil's agricultural sector features well within the top worldwide emitters of greenhouse gas, for sustainable practice, green bond funding is being taken up increasingly.
- **Multilateral Development Banks MDBs**: World Bank and IFC institutions have been issuing climate bonds, which spur growth specially in emerging markets, whose efforts helped create a local green market that will finance projects that improve climate resilience and enhance renewable energy infrastructure around countries like South Africa and Indonesia.

11.10 Future Prospects for Green and Climate Bonds

Green and climate bonds will certainly continue growing and changing to embrace novel challenges and circumstances, very open. As such, growth for the market is still projected as investors continue to demand more sustainable investments, and huge capital is required to help countries transition to a low-carbon or a climate-resilient world. More complex financial instruments, like sustainability-linked bonds and green securitizations, are quite likely to open up even more opportunities for investors and issuers alike. For instance, technological advance such as blockchain and advanced analytics of data will increase much transparency, impact measurement, and better market efficiency. Innovations, therefore, can further enhance the credibility of green bonds and push them into mainstream financial markets. In summary, green and climate bonds will become increasingly important in the mobilizing of capital for more sustainable development. Provided that the market overcomes regulatory challenges, scales up finance, and embraces innovation, these innovations will continue to facilitate meaningful environmental and social impact.

11.10.1 Challenges and Future Directions

Although green and climate bonds are gaining pace, it still has a long way forward with many challenges remaining. The major challenge lies in non-uniform definition of what constitutes a "green" project. International standards, such as the GBP, have promoted greater consistency but regional differences still remain, creating problems in making a proper comparison of bonds. The lack of liquidity in emerging markets, with issuances being less significant, is another challenge. The need to correct the imbalance has called for concerted efforts by both international organizations and governments toward a friendly environment for green bond issuance in developing economies. As we all know, in future, green and climate bonds will increasingly be a lifeline for financing the global low-carbon transition. Future governmental and policymaker activity in the process should consist of developing regulatory frameworks that give sustainable finance fertile ground to grow. Issuers, investors, and international bodies need coordination for the prolonged credibility and effectiveness of green bonds as instruments for climate action. In other words, Green and climate bonds have evolved from niche instruments to mainstream financial tools for addressing environmental challenges. The use of green bonds will play a crucial role in mobilizing much-needed capital toward sustainable projects toward a resilient and sustainable global economy. With further innovation, cooperation, and clearly defined policy frameworks, green and climate bonds will continue to remain at the center stage of activity in the field of sustainable finance, significantly contributing to efforts being made worldwide against climate change as well as environmental sustainability.

11.10.2 Research Findings

- **Background and Impact**: International standards in the development of green and climate bonds can be said to be related to two positive trends: the issues have now turned into critical instruments in financing projects aimed at overcoming climate and environmental challenges. These issuers and investors become involved ever more. Several green and climate bonds have demonstrated efficiency in financing renewable energy, energy efficiency, and various initiatives for climate resilience. With social benefits such as the provision of jobs and improvement in public health, their value extends beyond returns on investment.
- **Rules and Legal Framework**: The regulatory environment is heterogeneous, and therefore, there is no single one-region, one-size-fits-all standard and policy. While GBP and Climate Bonds Standards have brought much-needed guidance on issuance and reporting, fragmented regulations and, of course, policy gaps are sizeable challenges. Harmonization of these frameworks and resolution of conflicting policies constitutes an important step toward an integrated, hence a more efficient market.
- **Industry Barriers and Innovations**: Some of the important issues concerning green bonds' future include scaling up green finance and integrating it into policy gaps and innovative financial products. For instance, only transparency at markets will improve with the massive inclusion of more issuers and investors who, coupled with further teeming adoption of technology, advancement in Fin instrument, and data analytics, advance their markets and make green bonds very effective for achieving the set goals in environmental objectives.

Funding NA.

Conflict of Interest: There is no conflict of interest.

References

Bachelet MJ, Becchetti L, Manfredonia S (2019) The green bonds premium puzzle: the role of issuer characteristics and third-party verification. Sustainability 11:1098

Baker MP, Bergstresser DB, Serafeim G, Wurgler JA (2018) Financing the response to climate change: the pricing and ownership of U.S. green bonds. Working Paper

Baker M, Bergstresser D, Serafeim G, Wurgler J (2022) The pricing and ownership of US green bonds. Annu Rev Financ Econ 14(1):415–437

Ballester L, González-Urteaga A, Shen L (2024) Green bond issuance and credit risk: international evidence. J Int Finan Markets Inst Money 94(102013):102013. https://doi.org/10.1016/j.intfin.2024.102013

Bhutta US, Tariq A, Farrukh M, Raza A, Iqbal MK (2022) Green bonds for sustainable development: review of literature on development and impact of green bonds. Technol Forecast Soc Chang 175(121378):121378. https://doi.org/10.1016/j.techfore.2021.121378

Broadstock DC, Cheng LTW (2019) Time-varying relation between black and green bond price benchmarks: macroeconomic determinants for the first decade. Financ Res Lett 29:17–22. https://doi.org/10.1016/j.frl.2019.02.006

Cheong C, Choi J (2020) Green bonds: a survey. J Deriv Quant Stud 선물연구 28(4):175–189. https://doi.org/10.1108/jdqs-09-2020-0024

Cioli V, Giannozzi A, Roggi O, COLONNA LA (2021) Corporate green bond and stock price reaction. Int J Bus Manag 16, 0–0

Cortellini G, Panetta IC (2021) Green bond: a systematic literature review for future research agendas. J Risk Financ Manag 14(12):589. https://doi.org/10.3390/jrfm14120589

Dell'Atti S, Di Tommaso C, Pacelli V (2022) Sovereign green bond and country value and risk: evidence from European Union countries. J Int Financ Manag Acc 33(3):505–521

De Spiegeleer J, Schoutens W (2019) Sustainable capital instruments and their role in prudential policy: reverse green bonds. Working Paper

Fatica S, Panzica R, Rancan M (2021) The pricing of green bonds: are financial institutions special? J Financ Stab 54:100873

Flaherty M, Gevorkyan A, Radpour S, Semmler W (2017) Financing climate policies through climate bonds—a three stage model and empirics. Res Int Bus Financ 42:468–479. https://doi.org/10.1016/j.ribaf.2016.06.001

Flammer C (2020) Green bonds: effectiveness and implications for public policy. Environ Energy Policy Econ 1(1):95–128

Gianfrate G, Peri M (2019) "The green advantage: exploring the convenience of issuing green bonds". J Cleaner Prod 219 Forthcoming

Gianfrate G, Peri M (2019) The green advantage: exploring the convenience of issuing green bonds. J Clean Prod 219:127–135. https://doi.org/10.1016/j.jclepro.2019.02.022

Glomsrød S, Wei T (2018) Business as UNusual: the implications of fossil divestment and green bonds for financial flows, economic growth and energy market (February 16, 2016). Energy Sustain Dev 44:1–10. https://doi.org/10.1016/j.esd.2018.02.005

Krueger P, Sautner Z, Starks LT (2020) The importance of climate risks for institutional investors. Rev Fin Stud 33(3):1067–1111

Kumar R, Khanna R (2023) RPA (Robotic Process Automation) in finance & accounting and future scope. In: 2023 second international conference on augmented intelligence and sustainable systems (ICAISS). IEEE, pp 1640–1645

Kumar R, Kandpal B, Ahmad V (2023a) Industrial IoT (IIOT): security threats and countermeasures. In: 2023 international conference on innovative data communication technologies and application (ICIDCA). IEEE, pp 829–833

Kumar R, Malholtra RK, Grover CN (2023b) Review on artificial intelligence role in implementation of Goods and Services Tax (GST) and future scope. In: 2023 international conference on artificial intelligence and smart communication (AISC). IEEE, pp 348–351

Kumar R, Sexena A, Gehlot A (2023c) Artificial intelligence in smart education and futuristic challenges. In: 2023 international conference on disruptive technologies (ICDT). IEEE, pp 432–435

Kumar R, Lande A, Kumar D, Malhotra RK, Sharma A (2023d) Technology bridging-in entrepreneurs and consumers in product development. In: 2023 11th international conference on intelligent systems and embedded design (ISED). IEEE, pp 1–5

Kumar R, Saxena A, Singh R (2023e) Robotic process automation bridge-in banking institute and consumers. In: 2023 international conference on disruptive technologies (ICDT). IEEE, pp 428–431

Kumar V, Kumar R, Goel R, Singh T, Kaur MJ (2024a) The evolution of digital marketing technologies: a fuzzy TOPSIS approach to selected technology in E-commerce platform. Int J Syst Assur Eng Manag 1–14

Kumar R, Goel R, Singh T, Mohanty SM, Gupta D, Alkhayyat A, Khanna R (2024b) Sustainable finance factors in Indian economy: analysis on policy of climate change and energy sector. Fluctuation Noise Lett 23(2)

Lebelle M, Lajili Jarjir S, Sassi S (2020) Corporate green bond issuances: an international evidence. J Risk Fin Manag 13(2):25
Lebelle M, Lajili Jarjir S, Sassi S (2020) Corporate green bond issuances: an international evidence. J Risk Financ Manag 13(2):1–21
Logutov AV (2023) Global "green bond" market analysis and perspective of its development. In: Smart green innovations in industry 4.0 for climate change risk management. Springer International Publishing, Cham, pp 625–631
Maltais A, Nykvist B (2020) Understanding the role of green bonds in advancing sustainability. J Sustain Finance Investment 1–20
Otek Ntsama UY, Yan C, Nasiri A, Mbouombouo Mboungam AH (2021) Green bonds issuance: insights in low- and middle-income countries. Int J Corp Soc Responsib 6(1). https://doi.org/10.1186/s40991-020-00056-0
Park S (2018) Investors as regulators: green bonds and the governance challenges of the sustainable finance revolution. Stanford J Int Law 54(1):2018
Rümpf G (2019) A decade of green bonds–the origins, the present and the future
Stellner C, Klein C, Zwergel B (2015) Corporate social responsibility and Eurozone corporate bonds: the moderating role of country sustainability. J Bank Finance 59:538–549. https://doi.org/10.1016/j.jbankfin.2015.04.032
Tang DY, Zhang Y (2020) Do shareholders benefit from green bonds? J Corp Finan 61:101427
Wang KH, Zhao YX, Jiang CF, Li ZZ (2022) Does green finance inspire sustainable development? Evidence from a global perspective. Econ Anal Policy 75:412–426
Yeow KE, Ng SH (2021) The impact of green bonds on corporate environmental and financial performance. Manag Financ 47(10):1486–1510
Zerbib OD (2019) The effect of pro-environmental preferences on bond prices: evidence from green bonds. J Bank Finance 98:39–60. https://doi.org/10.1016/j.jbankfin.2018.10.012

Chapter 12
Mitigating the Climate Risk: The Role of Carbon Taxation

Radnyi Godase, Neeraj Gupta, and Prachi Chaturvedi

Abstract The carbon tax serves as a fiscal instrument designed to address climate change by levying a financial charge based on the carbon content of fossil fuels. The primary objective is to internalize the external costs associated with carbon emissions, thereby fostering the transition toward cleaner energy alternatives and enhancing energy efficiency. By raising the costs of carbon-intensive fuels, the carbon tax creates incentives for the reduction of greenhouse gas emissions, catalyzes advancements in environmentally sustainable technologies, and generates public revenue that can be allocated to support initiatives aimed at climate mitigation. This chapter offers an extensive examination of the carbon tax and its significance in confronting the multifaceted challenges posed by climate risk.

Keywords Carbon tax · Climate risk · Sustainability · Emission reduction · Fiscal instrument

12.1 Introduction

The carbon tax is a fiscal tool intended to combat climate change by imposing a financial penalty based on the amount of carbon present in fossil fuels. Internalizing the externalities linked to carbon emissions is the main goal in order to promote the switch to cleaner energy sources and improve energy efficiency. By increasing the financial consequences associated with carbon- intensive fuels, the carbon tax promotes greenhouse gas emission reduction, green technology innovation, and government revenue that can be used to fund climate mitigation initiatives (Sarpong et al. 2023).

R. Godase (✉)
School of Economics and Commerce, MIT World Peace University, Pune, India
e-mail: radnyi.p.godase@gmail.com

N. Gupta · P. Chaturvedi
N.L. Dalmia Institute of Management Studies and Research, Mumbai, India

R. Kumar (ed.), *Sustainable Finance*, Sustainability Solutions,
https://doi.org/10.1007/978-3-031-97962-0_12

The implications of carbon taxes are extensive and can affect each sector differently. From the enterprises' perspective, the production cost in manufacturing industries is relatively higher than that of the service industries, and they tend to recover either through increased prices or through improvement in efficiency. The transportation sector experiences increased fuel prices, further influencing operational strategies and purchase decisions. Consequently, higher energy costs force consumers to invest in energy-efficient alternatives affecting their lifestyles. However, these rising prices can, in turn, alter the general cost of living, especially for families with low income (Fremstad and Paul 2019). To counter these effects, policies, rebates, or subsidies are necessary.

A recent study conducted in China found that the carbon tax enhances air quality, which is good for human health (Li et al. 2018). Furthermore, in order to reduce energy intensity by 3–5% annually between 2015 and 2030, it takes into account different scenarios for carbon tax rates. By 2030, a carbon tax rate of US\$72/tCO_2 (2007 price) would be required to lower CO_2 intensity by 4% a year. When compared to the baseline (or no carbon tax) scenario, the carbon tax would result in a 24% reduction in US CO_2 emissions by 2030 (Timilsina 2018). As a way to decarbonize the global energy supply, carbon taxes are widely supported. However, some countries are using their energy policies to impose "negative" carbon taxes on fossil fuel consumers, i.e., lowering fossil fuel prices to practically zero.

The World Bank's report 'State and Trends of Carbon Pricing 2024' revealed that the revenues from carbon pricing had reached at historically high level of $104 billion in 2023. More than 50% of this revenue generated was used for climate related programs. The report emphasizes the importance of carbon pricing as a crucial tool to reduce carbon emissions.

In the initial report, the carbon taxes and emissions trading systems covered only about 7% of the world's emissions but in 2024, 24% of global emissions are covered. The report portrays that the middle-income countries with sizeable emissions, such as India, Brazil, and Türkiye, are moving toward implementing carbon pricing. Further, the EU Carbon Border Adjustment Mechanism pushes governments to consider carbon pricing for iron, steel, aluminum, cement, fertilizers, and energy industries. Global carbon price coverage and levels are still too low to meet the objectives of the Paris Agreement, despite record revenues and growth. Less than 1% of global greenhouse gas emissions are currently covered by direct carbon pricing, which is at or above the range established by the High-Level Commission on Carbon Prices to keep global warming below 2 °C (World Bank 2024).

12.1.1 Historical Context of Carbon Tax

Carbon taxation is a relatively new concept that emerged primarily due to growing concerns about climate change and impending action to address it. Initially, Finland became the first to introduce a carbon tax in 1990. With time, countries started adopting one or another form of carbon pricing, like cap-and-trade and carbon tax.

Thus, with the increasing recognition of climate change, carbon taxation has broadened its acceptance, moving from a handful of pioneering nations to a measure more broadly embraced by the public (Parry 2019; Wood 2018).

Countries like Sweden, the UK, Canada, and Australia have already woven carbon taxes or other such approaches to their environmental policies and adapted these taxes to fit the respective economic and environmental contexts (SBS News 2016). The evolving landscape of carbon taxes demonstrates, the increasing acceptance of market-based solutions concerning the issue of climate change. Moving toward these systems is also an implicit acknowledgment of the viability of market-based approaches for alleviating the impacts of climate change on a global scale.

12.1.2 Rationale Behind Carbon Tax

The carbon tax is a policy instrument designed to decrease CO_2 emissions and make using carbon-rich fuels more costly. It is a good deterrent against the use of fossil fuels and encourages the use of renewable energies and energy-efficient technologies by taxing the use of fossil fuels.

The carbon tax is supposed to internalize the external costs of carbon emissions into the firms' cost structures. That considers the effect on the environment and public health due to climate change, such that it accurately represents the "true" economic cost. Thus, promoting the creation of new technology because companies and industries can be motivated to reduce the costs of carbon-intensive activity in order to lower the tax burden (Sarpong et al. 2023). This encourages companies to research and utilize other forms of energy efficient technologies in order to avoid paying taxes (Kumar et al. 2024).

Also, carbon taxes provide the government with revenue that can be used to fund climate change projects, the development of renewable energy, and the subsidization of low or no carbon emission technologies. The revenue generated from carbon taxes can be used to offset other taxes or to subsidize low-income families affected by the increased cost of energy (Fremstad and Paul 2019).

At last, it provides a "positive feedback loop" in that it will encourage individuals and companies to change their behavior to conserve energy, be more energy efficient, and shift to less carbon-intensive goods and services, which will help in countering the global warming concern (Baker III et al. 2017).

12.1.3 Types of Carbon Tax Systems

1. Carbon Pricing: The umbrella term carbon pricing would apply to both forms of carbon taxes and cap-and-trade systems. Carbon taxes impose direct charges per unit of carbon emissions, while cap-and-trade systems create a market for permits to emit a given quantity of pollutant (Wood 2018).

2. Cap-and-Trade: Also referred to as emissions trading systems, cap-and-trade programs involve establishing the maximum level of emissions and creating a market of emissions permits. Such companies are free to sell their extra permits to other firms that cannot meet the emission limit set for them. The price of permits can be said to be somewhat like a carbon price, albeit in the form of a tax (Wood 2018).
3. Hybrid Systems: Certain countries adopt integrated arrangements of carbon taxes and cap-and-trade programs. These systems are intended to combine the advantages of use of both approaches in carbon pricing: flexibility and stability (Wood 2018).

12.2 Carbon Tax: Economic Consideration

12.2.1 Impact of the Different Industries

1. Energy Sector: The carbon taxes closely affect the energy sector because energy sources such as fossil fuels become more expensive. This may lead to higher costs for coal, oil, and natural gases in the market. Consequently, the energy companies might record increased operating expenses and lower profit margins. However, it also encourages capital investment in renewable technologies and clean energy as organizations seek to reduce their taxes (Macaluso et al. 2018).
2. Manufacturing Sector: By increasing the level of carbon tax, manufacturing-intensive industries like steel and cement producers are likely to face high production costs (Ganapati et al. 2020). This makes them prone to cost risks since they are usually associated with high carbon emissions. The added costs may be passed on to consumers through higher prices for goods, or companies may seek to improve energy efficiency and adopt cleaner technologies to offset the tax impact. The carbon tax may also affect the geographical location of new manufacturing facilities, thus creating pressure toward transferring manufacturing facilities to countries with lower carbon costs (Yang and Yan 2023).
3. Transportation Sector: Carbon taxes may increase fuel prices for industries involved in transporting goods through road, rail, air and waterways. This would imply an increase in operational costs in transportation companies, creating pressure on product and service prices. Organizations may purchase more efficient vehicles or use other fuels to minimize their tax burden. In addition, consumer behavior trends may change due to changing consumer preferences as consumers may be in search of the cheapest and most environmentally friendly means of transport (Köppl and Schratzenstaller 2023).
4. Agriculture and Forestry: Firms operating in the agriculture and forestry industry may also face increased costs due to fuel and energy use. Yet, these sectors can benefit from carbon tax policies and avail incentives for sustainable activities and carbon sequestration programs. For example, farmers may adopt low-emission

practices such as practicing no-till farming or reforestation to receive the linked subsidy or incentive of a carbon tax rebate (Gramig and Widmar 2018).

12.2.2 Cost Pass Through and Competitiveness

Firms might offset an increase in production costs arising from carbon taxation through price increase mechanisms that involve transferring this cost onto consumers in the market via various factors such as price elasticity, capacity costs, and competition in the market. While highly competitive industries may bear increased costs to maintain prices, firms with substantial pricing power or those in less competitive environments could pass on a higher portion of the costs to the consumer (Ganapati et al. 2020). The imposition of carbon taxes may induce market inefficiencies for domestic industries regarding global competitiveness. Firms in countries that are obliged to remit carbon taxes may find themselves incurring higher production costs than those located in countries that do not impose taxes on carbon. These situations pose a risk to climate initiatives through carbon leakage, which is when firms relocate to countries with lower carbon pricing (Yang and Yan 2023). In response to these pressures, policymakers could consider implementing carbon border adjustments or subsidies in specific sectors to maintain competitiveness.

12.2.3 Consumer Behavior

Higher carbon taxes may nudge consumers to adopt energy-efficient practices. For instance, people might invest in energy-efficient appliances, vehicles, and home insulation to reduce their carbon taxes. Efforts to transition to more efficient products and practices may be helpful in total emission reductions. Consumers can also consider purchasing renewable energy sources and using biofuels for transportation. As carbon taxes affect the cost of fossil fuels, renewables have a new demand because of cost changes. The carbon taxes' financial impact may entail actions like reduction in travel or, the use of more public transport or changes in diet (e.g., reduced meat consumption). These changes result from the need to reduce individual carbon footprint and the effect that high energy costs will have (Parry 2019).

In principle, carbon taxes are mostly followed by an increase in the prices of goods or services with notable carbon footprints. This can alter the cost of living because the consumers utilize carbon-intensive services such as energy, transport, etc. This means that the degree of the price increase depends on the carbon tax rate, the extent of cost transfer, and its interconnections with the supply chain. The effect of higher prices on the standard of living will also differ from one income group to another. The rise in essential services such as energy or transportation costs may disproportionately affect lower-income households (Fremstad and Paul 2019), sparking debates over whether the 'poor' should receive rebates or subsidies.

12.3 Revenue Generation and Use

12.3.1 How Carbon Tax Revenues Are Utilized

The government may receive the revenue generated by the carbon tax directly. Therefore, this revenue could be utilized to support renewable energy, enhance public transportation, and mitigate and adapt to climate change. The profits generated may also lessen the impact on businesses and private entities by compensating for the reduction of other taxes, such as income or corporate tax. Innovations in power generation, public transportation, public buildings, and other climate-related infrastructures can all be utilized to improve the efficiency of public service delivery through the carbon tax. These investments can support the broader climate goals while lessening the impact of carbon taxes on the private sector and consumers (Köppl and Schratzenstaller 2023; Mintz-Woo 2024).

12.3.2 Potential for Revenue Recycling and Investment in Green Technologies

Carbon tax revenue shall be used through "revenue recycling", whereby the money raised through carbon taxes will be paid back into the economy by either lower taxes or direct rebates. For example, some governments reduce corporate taxes and income taxes with carbon tax revenues or giving rebate to households. Recycling tax proceeds should make the tax less burdensome economically, bolster support, and widen acceptance of carbon pricing policies (Beck et al. 2015; Beiser-McGrath and Bernauer 2019).

The third significant way is to invest into green technologies and sustainable infrastructure. Governments can use the proceeds raised from the carbon tax to research and develop clean energy technologies, support renewable energy systems, and promote energy conservation programs. Investment in green technologies can stimulate innovation, create jobs, and spur the transition to a low-carbon economy (Jabeen et al. 2025).

12.4 The Rising Climate Risk

Climate risk refers to type of risks that consist of potential operational and financial effects arising from climate-related factors such as extreme weather conditions and regulatory actions aimed at reducing climate change effects. The impacts of climate risk are primarily on businesses and the economy at large. Disruptions caused as a result of climate change often come at increased operating costs for businesses and may expose to further risks that arise from changing regulations such as carbon pricing. The absence of timely adaptation involves the threat of decreased competitiveness and further lowers the reputation of the firms alongside higher costs of compliance (Kumar et al. 2024).

At a macroeconomic level, climate risk can impede economic stability by way of supply chain interruptions, fluctuations in commodity prices, changes in investment patterns. Examples include fossil fuel-reliant industries facing higher costs from carbon taxes and industries investing in renewable energies benefitting from subsidy policies (Campiglio et al. 2023).

12.4.1 *Carbon Tax and Climate Risk Management*

The carbon tax mechanism creates a business incentive for enterprises to factor climate risk into their decision-making related to risk management (Li et al. 2022a, b) along with how efficiently and sustainably resources are used, while maximally exploiting renewable alternatives and creating more integration among supply chains. With proactive approach, firms will be able to show their leadership in avoiding financial risk but also contribute significantly to climate risk management. In addition, carbon taxes require firms to start looking at climate risk management from lens of strategic planning. Therefore, firms will be obliged to examine in great detail as to how climate change may impact their business operations, supply chains, and the market environment in molding resilient carbon pricing and climate risk strategies (Mcfarland et al. 2018). This fosters the overall corporate sustainability. Besides, incorporating developments in carbon tax, the involvement tends to compel businesses to become transparent and accountable for their emissions and reporting.

Thus, improved environmental reporting practices enhanced with the accurate account of carbon emissions and carbon-tax obligations confer a better way for the stakeholders to build trust, sign compliance under the law, and derive valuable insights into their climate risk exposure (Macaluso et al. 2018). Introducing carbon taxes advocates a systemic approach toward sustainability and risk management in the corporate landscape.

12.4.2 *Synergies with Other Climate Policies*

Energy efficiency requirements, renewable energy mandates, and emissions trading programs are a few examples of climate policies that carbon taxes can support. Governments can boost the effectiveness of climate policies by creating a single regulatory framework. For example, a carbon tax could be used to complement cap-and-trade schemes by imposing a price on carbon emissions and permitting the market trading of emission allowances.

Coordinating the implementation of a carbon tax with another climate policy is necessary to ensure that the efforts to combat climate change complement and reinforce one another. Comprehensive climate changes, such as adhering to greenhouse gas targets and transitioning to a low-carbon economy, can be accomplished through policy synergies. Coordinated efforts reduce the possibility of policy conflict and improve climate policy's overall efficacy.

Integrating carbon taxes with R&D-supporting policies, subsidies for green technologies, and public investment in sustainable infrastructure can stimulate innovation and rapid transition to a low-carbon economy. For instance, revenues from a carbon tax can be invested in clean energy R&D, while renewable energy mandates secure the marketplace for these technologies (Barata et al. 2024).

12.4.3 Assessment and Measurement of Climate Risk

In the realm of climate change adaptation, various analytical methods have been developed to help organizations anticipate and mitigate the potential impacts of climate change on their operations. Numerous analytical methodologies have been designed to support organizations in predicting and reducing possibly adverse impacts of climate on their operations. Scenario analysis is one such method that hypothesizes potential future and investigates them to see how climate change-induced changes might impact business and financial performances (Macaluso et al. 2018). By navigating through some of the possible outcomes, organizations learn about the specific risks posed by climate variations and how to deal with them. It can encompass an evaluation of the potential impact of carbon taxes, analysis of severe weather events, and longer-term climate trends.

Second, carbon footprint tracking determines the level of greenhouse gas emissions an organization causes across its core functions. Having an insight into the source of an organization's greenhouse emissions helps to provide foresight knowledge on activities that could neutralize emissions causing GHG (McAusland and Najjar 2015). Also, it is a proper tool or means to help the organizations analyze how much exposure to carbon taxes the organization is being exposed to and also provide the organization with an available method of managing or mitigating the risks.

Climate risk modeling is another tool that uses statistical methods and data to estimate an organization's possible operational and financial impact of climate change on the company, including carbon footprint tracking and other climate factors (Smith et al. 2015). These models guide an organization on deciding the resource allocation or expenditure concerning climate risk management. Finally, Environmental Impact Assessments (EIAs) scrutinize ongoing and potential environmental impacts of business activities and projects. This indicator will analyze how carbon taxes and other climate policies affect environmental performance. The EIAs, in particular, help companies understand regulatory requirements and how they may integrate climate risk into project planning and decision-making processes (Yang et al. 2014).

12.5 Policy Design and Implementation

12.5.1 Key Considerations for Effective Carbon Tax Policy

The policymakers working toward carbon tax should pursue specified objectives, such as lowering emissions, promoting clean technologies, or generating revenues for climate-related investments. The policy framework is determined by clearly defined objectives which will also facilitate policy evaluation. It must also be set reasonable level to influence any changes in behavior that will meet emissions targets with minimal effect on the economy. It should also be adjusted periodically depending on information regarding climate change, technological advancement, and on the economic environment.

From large industries to transportation and energy production, a carbon tax should ideally be levied on a comprehensive selection of sources. Its extensiveness alone assures emissions impact and curbs loopholes that might arise.

The success of carbon tax policy lies in its stability and predictability, for businesses and investors as they draw comfort from that predictability. The sustainability of the policy helps investors make informed choices about their future investments in low-carbon technologies.

A stable revenue allocation plan through which the funds generated through the carbon tax will be redistributed is another critical feature of an effective carbon tax. Possible options include financing renewable energy, subsidizing energy conservation initiatives, cutting other taxes, or issuing rebates to affected families. Indeed, how the revenues are used will enhance the attractiveness and efficiency of a scheme.

12.5.2 Ensuring Fairness and Equity

Households with lower incomes may feel the most adverse effects of carbon taxes compared to richer ones. This is because poorer households use a larger share of their income for energy and transportation costs. To address this, rebates or subsidies could be suggested where policymakers care for the at-risk populations.

Specific energy-intensive industries may have serious concerns imposed by the new carbon tax. Hence, it is essential that relief measures are included in the policies and implemented at any time to ease some transition burdens on them.

Carbon tax-related policies should also consider regional differences, such as the state of the economy, energy sources and emission profiles in different regions. Such a tailored approach promotes fairness and efficiency by considering the peculiar necessities of the region. For example, regions that depend heavily on fossil fuels will need targeted intervention to steer them toward using low-carbon alternatives.

Furthermore, in implementing a carbon tax policy, consideration must also be made regarding global competitiveness and carbon leakage, as these may drive businesses away from areas with severe regulations on emissions. A border adjustment mechanism would thus help guarantee that the competitive position of domestic industries is not impaired as they compete in the world market.

12.6 Implementation Challenges

12.6.1 Administrative and Regulatory Issues

Management of carbon taxes necessitates complex administrative tasks like emissions monitoring, tax management, and compliance. For example, the government need to have adequate structures and facilities in place to perform these activities efficiently and to minimize tax evasion risks.

Above several aforementioned processes of tax implementation are some other required considerations. Carbon tax policies need to harmonize with other environmental policies and standards such as emissions trading schemes, energy efficiency standards, and renewable energy requirements, to avoid conflicting outcome and ensure coherent regulatory framework.

Emissions assessment and calculation of tax liabilities require accurate and verified data. Global administrators and corporate entities must ensure that they deploy effective data management, and verification mechanisms to support the reliability of the emission reporting systems.

12.6.2 Addressing Industry Resistance and Public Perception

Specific industries may resist carbon taxes as they will be paying more for their inputs and can lobby for exclusion or lower tax rates. Thus, the government needs to address the industry's concerns and demonstrate the long-term benefits of the policy to gain support and reduce resistance.

While people are aware of carbon taxes, they can have particular concerns about the definite increase in living costs or other economic consequences. Effective policy dissemination and timely consultations can help build public support and address misconceptions (Zhang and Baranzini 2004).

Dissemination of benefits to the businesses and the public, such as environmental conservation, enhanced public health and long-term economic gains, can foster better acceptance and support. It is also essential to define the positive aspects, explain them to the target audience and address possible concerns.

12.7 Monitoring and Evaluation

12.7.1 Metrics for Assessing Policy Effectiveness

Carbon tax policy can be assessed in terms of the effectiveness of specific predetermined indicators; the primary one is a reduction in greenhouse gas emissions. Emissions data will track whether the policy achieves its purpose of reducing emissions successfully (Bruvoll and Larsen 2017). Meanwhile, other aspects of the carbon tax's economic influence must be studied: these include energy prices, business costs, and economic growth. The economic dimensions of the policy are reflected in continuous measures including GDP growth, sectoral impacts like carbon emissions, and inflation as reflected in the consumer price index. Also, it is vital to also assess the effects of the carbon tax on innovation, which includes tracking investments in clean technology and energy efficiency. The key metrics are R&D expenditures, renewable energy adoption, and the development of low-carbon technologies, all establishing themselves as synergies on which innovation policy success is assessed (Gerlagh et al. 2009). Finally, an overall evaluation of carbon tax policy must take into account what the tax revenue will be spent on. These include assessing the proportional allocation of funds to climate projects, by reducing other taxes, and through rebates, which are essential in attaining the objectives of the policy (Beck et al. 2015; Beiser-McGrath and Bernauer 2019). The trade-off between these various measures allows the carbon tax policy to pan out its efficiency.

12.7.2 Adjustments and Policy Refinement

Regular assessment with respect to carbon tax policies must assume primary importance to study the effectiveness and adapting it to the evolving environmental and economic conditions. Review allows changes in the rate of taxation to broaden the coverage and handle complications arising due to uncertainties. Such periodic reviews create opportunities for policymakers to respond to technological changes concerning climate or economic conditions. Stakeholder feedback is a critical intervention to improve. Consultations and surveys of these involved can offer policymakers insights into the effects of the carbon tax and enable targeted changes (Dhanda et al. 2022). To some extent, the case studies of how other countries or regions have approached similar initiatives can provide valuable lessons.

12.8 Emerging Carbon Tax Models

12.8.1 Innovations in Carbon Pricing Mechanisms

1. Hybrid Carbon Pricing Models: These emerging hybrid models integrate characteristics of carbon taxes and cap and trade schemes for their complementary advantages. A hybrid scheme may set a carbon tax to create a price floor, and a cap-and-trade system works to set price ceilings. Price stability and flexibility can encourage emission-reducing actions, while minimizing economic disruptions (Quere et al. 2016).
2. Carbon Tax with Border Adjustments: In a bid to tackle concerns of carbon leakage, i.e., when companies may shift to countries where standards are negligible, some recent strategies propose border adjustment measures. Import penalties and export rebates are some mechanisms can be considered to maintain competitiveness against conditions created by lower carbon prices (Kortum and Weisbach 2017). This approach is to discourage the exodus of the right companies and to protect domestic industries.
3. Dynamic Carbon Pricing: Among the innovations in carbon pricing is a dynamic or variable pricing model with real-time changes adapting to the most recent emissions output, market trends, or available technologies (Gambardella et al. 2020). Such dynamic pricing has the potential to enhance carbon tax effectiveness by responding to changing conditions, in order to keep the tax rate effective in engaging in driving lower emissions.
4. Sectoral Carbon Taxes: Such taxes may take on specific concerns of individual sectors such as transportation, agriculture, or heavy industry (King et al. 2019). If tax is made appropriate to the sensitivities of individual sectors, it will enhance the effectiveness and fairness.

12.8.2 Integration with Global Carbon Markets

1. Linking National and Regional Carbon Markets: Integrated national and regional carbon markets can produce a larger, more efficient carbon trading market (Woerman 2023). Through market linkage, countries can pool resources to share the emission reductions and obtain the lowest possible cost. For example, The European Union Emissions Trading System (EU ETS) is associated with other regional emission trading system, enhancing cross border emission market.
2. Global Carbon Pricing Initiatives: The focus has been shifting to creating international carbon pricing programs that harmonize national goals with global goals. Organizations like the World Bank and the International Carbon Action Partnership (ICAP) are currently developing frameworks to help facilitate the convergence of carbon pricing around the globe (Wettestad et al. 2021). Such initiatives have been made with the overall goal of improving the compatibility of the carbon

pricing system at the international level to minimize the likelihood of the market becoming fragmented and to increase the efficiency of carbon pricing initiatives worldwide.

3. Carbon Pricing and Trade Agreements: Future trends include the use of carbon pricing instruments as a tool of policy in international relations. This could entail creating links between carbon pricing provisions of the trade liberalization and climate change accord and harmonizing trade policies with carbon reduction objectives. Thus, integration can facilitate international cooperation on climate policies and address competitive concerns related to carbon pricing.

12.9 Technological Advancements

12.9.1 Impact of Technology on Carbon Tax Effectiveness

1. Enhanced Monitoring and Verification: Technological improvements, such as satellite monitoring facilitate emissions tracking and reporting by enhancing their accuracy and transparency. Monitoring technologies ensure accurate measurement of emissions, thus helping improve the effectiveness of carbon tax policies (Hu et al. 2024).
2. Data Analytics and A.I.: The role of data analytics and artificial intelligence can optimize the implementation and enforcement of carbon taxes. These A.I. models are capable of forecasting emission trends, reviewing policy impacts, and identifying areas of improvement. Analysis of the impact of carbon pricing may provide direction for consumer and business response.
3. Carbon Trading on Digital Platforms: The implementation of carbon trading and management of carbon credits can be made possible through Web-based solutions and blockchain. Compared to conventional methods of trading carbon allowances and confirming reductions in emissions, these technologies are secure, transparent, and efficient, thereby increasing the utility of carbon markets and the implementation of carbon taxes (Pan et al. 2019).

12.9.2 Role of Innovation in Reducing Carbon Emissions

1. Clean Energy Technologies: From solar and wind to advanced battery storage, the innovative technologies will provide significant support for carbon emissions reduction; funds from carbon tax revenues and complementary policy support, will fuel the transition to a low carbon energy system (Jabeen et al. 2025).
2. Energy Efficiency Improvements: While new technologies, such as smart grids, energy-efficient appliances, and industrial processes, will further help reduce energy consumption and emissions, a carbon tax could create an incentive for

investment in energy-efficient technologies and thereby stimulate their adoption, resulting in further cuts in emissions (Macaluso et al. 2018).
3. Carbon capture and storage (CCS): The revolutionary technologies of CCS offer measurable solutions for emission mitigation from industrial processes and power generation. Carbon taxes can drive investment in CCS technologies by making it more expensive to emit carbon, creating a market for its capture and storage solution (Bui et al. 2018).
4. Electrification of Transportation: Advances in electric vehicle technology and associated infrastructure are reshaping the automotive sector; a carbon tax can increase the cost of fossil-fuel vehicles, making the electrification of transportation a much more competitive avenue for a cleaner transportation system (Nadel 2019).

12.10 Global and Regional Developments

12.10.1 Trends in International Carbon Tax Policies

In the last few years, several countries have adopted schemes for carbon taxes as one of their climate action strategies. Specifically, Canada and Japan are said to be reshaping their carbon taxes in the face of steep post-COP 21 targets (Pan et al. 2024). Other initiatives, such as the Carbon Pricing Coalition (CPLC) and the Global Carbon Pricing Challenge assist in platforming dialogue and coalescence with other international cooperations toward carbon-pricing mechanisms (Wettestad et al. 2021). Under carbon taxes, a wide ambit of formal commitment exists, which becomes legally enforceable under international agreements and frameworks. One such opportunity exists through the Paris Agreement, which promotes the use of carbon pricing by countries to achieve their Nationally Determined Contributions (NDCs) (Vogt-Schilb and Hallegatte 2017). The modality of international agreements integrates national policies to adhere the established climate goals and reinforce actions taken against climate change.

12.10.2 Prospects for Global Carbon Tax Agreements

There are ongoing discussions concerning the derivation of a coherent carbon pricing mechanism under the established Global Carbon Pricing Framework where common standards and guidelines for carbon pricing policies would be defined. Such a framework could inculcate an incentive for the member countries to synchronize their policies in order to reduce market fragmentation and allow external coordination and cooperation in global carbon pricing (Hussein and Khan 2023).

The possible linkage with other regional markets opens the possibility for a more integrated global carbon market—a global accord that might provide higher market liquidity, lower prices, and improve efficiency in carbon pricing, thus allowing for relatively cost effective emission reductions (Woerman 2023).

Future development could see international carbon tax agreements. They may purport to resolve the issues of carbon leakage, competitiveness, and equity from an international framework to cooperate in implementing carbon pricing. This accord will require tremendous degrees of international cooperation and negotiation, but if attained will provide enormous benefits in terms of emission reductions and climate action (World Trade Organisation 2022).

12.11 Case Studies

12.11.1 Examples of Companies Adapting to Carbon Tax

1. Microsoft: Microsoft imposes carbon tax on business units within the company for emissions produced. The revenue that is generated from the internal carbon tax is reinvested in specified sustainable initiatives, such as energy efficiency and renewable energy purchases. The effects of this approach have provided large reductions in the carbon footprint and established the company as one of the leaders in corporate sustainability (Light 2015).
2. Shell: The Shell Company has integrated carbon pricing into investment decisions and project evaluation processes, using internal carbon pricing to assess the viability of new projects and investments. By integrating carbon pricing into decision-making, Shell aims to align their investments in favor of long-term climate targets and financial risks related to carbon emissions (Li et al. 2022a, b).

12.11.2 National Carbon Tax Implementations: Detailed Analysis of Specific Countries

1. Sweden

Sweden's carbon tax, introduced in 1991, is one of the earliest known to be imposed. The tax applies to fossil fuel use in heating and transportation. The tax, under a relatively low rate (EUR 25), has gradually risen (EUR 125). In 2023, the carbon tax rate in Sweden was one of the highest in the world and symbolizes Sweden's serious engagements in reducing its emissions of greenhouse gases. Sweden has seen vast reductions in carbon emissions, particularly within the energy sector. The tax has been responsible for the very shift to renewable energy sources and greater energy efficiency. Economic growth has followed in the wake of these developments, which proves that a high carbon tax can exist in harmony with a strong economy (The Government Offices of Sweden 2024).

2. Canada

Canada imposed a federal carbon tax in 2019 under the Greenhouse Gas Pollution Pricing Act. It is imposed in provinces or territories with no carbon pricing facility of their own. The federal tax sets prices for emissions from fossil fuels, which they plan to increase over time. Provinces and territories may create their own carbon pricing systems as long as they meet federal standards. Initial findings suggest that emissions have largely fallen in provinces subject to carbon pricing, yet the carbon tax has faced political backlash and legal challenges. Simultaneously, it spurred investments in clean technology and renewable energy (Beck et al. 2015).

3. France

The carbon tax was introduced in France in 2014; it originally focused on fossil fuels used in power and transport. This provided an integral aspect of the French climate strategy in the context of its broader mitigation mission. Some adjustments in French carbon tax policy stem from public resistance and protests. Despite these challenges, the tax has impacted relatively lower fossil fuel consumption and created debates over the roles carbon pricing should play in meeting climate objectives. This experience from France has illustrated the major roles of public acceptance and a fair policy design (Fabre and Douenne 2022).

4. China

Before creating a nationwide carbon market in China in 2021, pilot emissions trading schemes (ETS) helped China in launching carbon pricing. Initially, the national ETS encompasses the power-generating industry and plans to extend to other sectors subsequently. The pricing mechanism for the carbon market is based on cap-and-trade rather than a carbon tax. Through the intermediate carbon market, China has a market-driven approach that should catalyze emissions reduction. Early indicators show that the ETS has helped in curbing emissions growth and further investment in cleaner technologies (Deng et al. 2018).

12.12 Conclusion

In conclusion, carbon taxes are effective in combating climate change by creating incentives for reduction in greenhouse gas emissions and support a shift toward cleaner energy sources. The economic impacts of carbon taxes differ among sectors and consumer decisions. However, when combined with strategic utilizations of these taxes, they can create positive changes in the environmental and economic balance. Businesses must integrate carbon pricing into their strategic planning as well as drive innovation. Policymakers must design carbon taxes with principles of equity and fairness, engaged the public as well as develop a comprehensive climate policy for enhanced effectiveness.

Funding NA.

Conflict of Interest There is no conflict of interest.

References

Baker III JA, Feldstein M, Halstead T, Mankiw NG, Paulson Jr. HM, Shultz GP, Stephenson T, Walton R (2017) The conservative case for carbon dividends. Climate Leadership Council. https://clcouncil.org/report/the-conservative-case-carbon-dividends/

Barata PM, Mukherjee M, Silitonga A, Jones D, Jaspal M, Hanafi A, Pearen H (2024) Navigating carbon pricing: the G20 experience and global south prospects. https://coilink.org/20.500.12592/fttf44r

Beck M, Rivers N, Wigle R, Yonezawa H (2015) Carbon tax and revenue recycling: impacts on households in British Columbia. Resour Energy Econ 41:40–69. https://doi.org/10.1016/j.reseneeco.2015.04.005

Beiser-McGrath LF, Bernauer T (2019) Could revenue recycling make effective carbon taxation politically feasible? Sci Adv 5(9):eaax3323. https://doi.org/10.1126/sciadv.aax3323

Bruvoll A, Larsen BM (2017) Greenhouse gas emissions in Norway: Do carbon taxes work? In: Environmental taxation in practice. Routledge, pp 545–557

Bui M, Adjiman CS, Bardow A, Anthony EJ, Boston A, Brown S, Fennell PS, Fuss S, Galindo A, Hackett LA (2018) Carbon capture and storage (CCS): the way forward. Energy Environ Sci 11(5):1062–1176

Campiglio E, Daumas L, Monnin P, von Jagow A (2023) Climate-related risks in financial assets. J Econ Surv 37(3):950–992. https://doi.org/10.1111/joes.12525

Deng Z, Li D, Pang T, Duan M (2018) Effectiveness of pilot carbon emissions trading systems in China. Climate Policy 18(8):992–1011. https://doi.org/10.1080/14693062.2018.1438245

Dhanda KK, Sarkis J, Dhavale DG (2022) Institutional and stakeholder effects on carbon mitigation strategies. Bus Strateg Environ 31(3):782–795

Fabre A, Douenne T (2022) Public support for carbon taxation: lessons from France. CEPR. https://cepr.org/voxeu/columns/public-support-carbon-taxation-lessons-france

Factbox: Carbon taxes and emission trading schemes around the world (2016). SBS News. https://www.sbs.com.au/news/article/factbox-carbon-taxes-and-emission-trading-schemes-around-the-world/2k5n0ufgi

Fremstad A, Paul M (2019) The impact of a carbon tax on inequality. Ecol Econ 163:88–97. https://doi.org/10.1016/j.ecolecon.2019.04.016

Gambardella C, Pahle M, Schill W-P (2020) Do benefits from dynamic tariffing rise? Welfare effects of real-time retail pricing under carbon taxation and variable renewable electricity supply. Environ Resource Econ 75:183–213

Ganapati S, Shapiro JS, Walker R (2020) Energy cost pass-through in US manufacturing: estimates and implications for carbon taxes. Am Econ J Appl Econ 12(2):303–342. https://doi.org/10.1257/app.20180474

Gerlagh R, Kverndokk S, Rosendahl KE (2009) Optimal timing of climate change policy: interaction between carbon taxes and innovation externalities. Environ Resource Econ 43(3):369–390. https://doi.org/10.1007/s10640-009-9271-y

Gramig BM, Widmar NJO (2018) Farmer preferences for agricultural soil carbon sequestration schemes. Appl Econ Perspect Policy 40(3):502–521. https://doi.org/10.1093/aepp/ppx041

Hu K, Zhang Q, Gong S, Zhang F, Weng L, Jiang S, Xia M (2024) A review of anthropogenic ground-level carbon emissions based on satellite data. IEEE J Sel Topics Appl Earth Observ Remote Sens 17:8339–8357. https://doi.org/10.1109/JSTARS.2024.3355549

Hussein ZRA, Khan FI (2023) The case for a global carbon-pricing framework. Foreign affairs. https://www.foreignaffairs.com/world/case-global-carbon-pricing-framework
Jabeen G, Wang D, Pinzón S, Işık C, Ahmad M, Rehman A, Anser MK (2025) Promoting green taxation and sustainable energy transition for low-carbon development. Geosci Front 16(1):101928. https://doi.org/10.1016/j.gsf.2024.101928
King M, Tarbush B, Teytelboym A (2019) Targeted carbon tax reforms. Eur Econ Rev 119:526–547
Köppl A, Schratzenstaller M (2023) Carbon taxation: a review of the empirical literature. J Econ Surv 37(4):1353–1388. https://doi.org/10.1111/joes.12531
Kortum S, Weisbach D (2017) The design of border adjustments for carbon prices. Nat Tax J 70(2):421–446
Kumar R, Goel R, Singh T, Mohanty SM, Gupta D, Alkhayyat A, Khanna R (2024) Sustainable finance factors in indian economy: analysis on policy of climate change and energy sector. Fluctuation Noise Lett 23(2)
Li M, Zhang D, Li C-T, Mulvaney KM, Selin NE, Karplus VJ (2018) Air quality co-benefits of carbon pricing in China. Nat Clim Chang 8(5):398–403
Li M, Trencher G, Asuka J (2022) The clean energy claims of BP, Chevron, ExxonMobil and Shell: a mismatch between discourse, actions and investments. PLoS ONE 17(2):e0263596
Li S, Wang H, Liu X (2022) The impact of carbon tax on financial stability. Environ Sci Pollut Res 29(37):55596–55608. https://doi.org/10.1007/s11356-022-19689-8
Light SE (2015) The new insider trading: environmental markets within the firm. Stanford Environ Law J 34(1):3–55
Macaluso N, Tuladhar S, Woollacott J, McFarland JR, Creason J, Cole J (2018) The impact of carbon taxation and revenue recycling on U.S. industries. Clim Change Econ 9(1). https://doi.org/10.1142/S2010007818400055
McAusland C, Najjar N (2015) Carbon footprint taxes. Environ Resource Econ 61(1):37–70. https://doi.org/10.1007/s10640-013-9749-5
Mcfarland JR, Fawcett AA, Morris AC, Reilly JM, Wilcoxen PJ (2018) Overview of the EMF 32 study on U.S. carbon tax scenarios. Climate Change Econ 09(01), 1840002. https://doi.org/10.1142/S201000781840002X
Mintz-Woo K (2024) Carbon tax ethics. Wires Clim Change 15(1):e858. https://doi.org/10.1002/wcc.858
Nadel S (2019) Electrification in the transportation, buildings, and industrial sectors: a review of opportunities, barriers, and policies. Curr Sustain/Renew Energy Rep 6(4):158–168. https://doi.org/10.1007/s40518-019-00138-z
Pan Y, Zhang X, Wang Y, Yan J, Zhou S, Li G, Bao J (2019) Application of blockchain in carbon trading. Energy Procedia 158:4286–4291
Pan J, Cross JL, Zou X, Zhang B (2024) To tax or to trade? A global review of carbon emissions reduction strategies. Energ Strat Rev 55:101508. https://doi.org/10.1016/j.esr.2024.101508
Parry I (2019) Back to basics: what is carbon taxation?—IMF F&D. https://www.imf.org/en/Publications/fandd/issues/2019/06/what-is-carbon-taxation-basics
Quere S, Delmas-Patterson G, Berezin A (2016) A new paradigm of governance for a carbon-pricing system. Int J Energy Prod Manag 1(2):192–205. https://doi.org/10.2495/EQ-V1-N2-192-205
Sarpong KA, Xu W, Gyamfi BA, Ofori EK (2023) Can environmental taxes and green-energy offer carbon-free E7 economies? An empirical analysis in the framework of COP-26. Environ Sci Pollut Res 30(18):51726–51739. https://doi.org/10.1007/s11356-023-25904-x
Smith BA, Ruthman T, Sparling E, Auld H, Comer N, Young I, Lammerding AM, Fazil A (2015) A risk modeling framework to evaluate the impacts of climate change and adaptation on food and water safety. Food Res Int 68:78–85. https://doi.org/10.1016/j.foodres.2014.07.006
The Government Offices of Sweden (2024) Sweden's carbon tax [Government]. https://www.government.se/government-policy/taxes-and-tariffs/swedens-carbon-tax/
Timilsina GR (2018) Where Is the carbon tax after thirty years of research? World Bank, Washington, DC. https://doi.org/10.1596/1813-9450-8493

Vogt-Schilb A, Hallegatte S (2017) Climate policies and nationally determined contributions: reconciling the needed ambition with the political economy. Wiley Interdisciplinary Reviews: Energy and Environment 6(6):e256

Wettestad J, Gulbrandsen LH, Andresen S (2021) Calling in the heavyweights: Why the World Bank established the carbon pricing leadership coalition, and what it might achieve. Int Stud Perspect 22(2):201–217

Woerman M (2023) Linking carbon markets with different initial conditions. J Environ Econ Manag 119:102820. https://doi.org/10.1016/j.jeem.2023.102820

Wood J (2018) The pros and cons of carbon taxes and cap-and-trade systems. Sch Public Policy Publ 11. https://doi.org/10.11575/sppp.v11i0.52974

World Bank (2024) State and trends of carbon pricing 2024. Washington, DC: World Bank. https://doi.org/10.1596/978-1-4648-2127-1. License: Creative Commons Attribution CC BY 3.0 IGO

World Trade Report (2022) Climate change and international trade. World Trade Organisation. https://www.wto.org/english/res_e/publications_e/wtr22_e.htm

Yang C, Yan X (2023) Impact of carbon tariffs on price competitiveness in the era of global value chain. Appl Energy 336:120805. https://doi.org/10.1016/j.apenergy.2023.120805

Yang XJT, Yang BLG, Lv H, Han WH (2014) An economic and environmental impact analysis of carbon tax policies: a case from inner Mongolia. Adv Mater Res 962–965:1701–1708. https://doi.org/10.4028/www.scientific.net/AMR.962-965.1701

Zhang Z, Baranzini A (2004) What do we know about carbon taxes? an inquiry into their impacts on competitiveness and distribution of income. Energy Policy 32(4):507–518. https://doi.org/10.1016/S0301-4215(03)00152-6

Chapter 13
Sustainable Finance: Strategies and Tools to Manage Climate Risk

Mamta Chawla, Malay Mohan Mahara, Amit Verma, Aparajita DasGupta Amist, and Garima Agarwal

Abstract This chapter will be exploring as to how smart technology investments go hand in hand with sustainability through technologies in India such as unified payment interface or UPI that help us achieve goals under both the environmental and economic constraints. With the global concerns regarding climate change, resource depletion and over exploitation of its wealth continue to rise, so does the need to look into improving the resource efficiency by the means of smart technologies in the sectors of Agriculture, Energy and urban development. Chapter puts across the significance of these technologies like IoT, AI and machine learning in terms of real-time data gathering and analysis that enable improvement in resource management. For instance, IoT sensors in agriculture help in the reduction of usage of water. India's UPI is shown as a revolutionary digital payment system that has simplified access to financial services that has helped in allowing financial inclusion for millions of people. This move from cash to digital payments has decreased environmental consequences by cutting the use of paper and metal, and lowering energy consumption related with currency processing. Despite these developments, concerns such as high implementation costs, infrastructure needs and cybersecurity threats are still to be addressed. To address such obstacles, the chapter suggests a multi

M. Chawla · A. D. Amist
Amity Global Business School, Noida, Amity University Uttar Pradesh, Noida, Uttar Pradesh, India
e-mail: adgupta@amity.edu

M. Chawla (✉) · A. Verma
Department of Commerce, Kalicharan PG College, Lucknow, Uttar Pradesh, India
e-mail: mchawla@amity.edu

A. Verma
e-mail: amitvm2007@gmail.com

M. M. Mahara
International Institute of Professional Studies, Devi Ahilya Vishwavidyalaya, Indore, Madhya Pradesh, India

G. Agarwal
Institute Amity Centre for Entrepreneurship Development, Amity University Uttar Pradesh, Noida, India

R. Kumar (ed.), *Sustainable Finance*, Sustainability Solutions,
https://doi.org/10.1007/978-3-031-97962-0_13

stakeholder strategy that highlights and supports government policies, public private partnerships, and international collaborations. The chapter finishes by pointing out the need for a balance between technical progress and environmental care. It asks for continuous and long-term study to improve the accessibility and environmental sustainability of smart technologies throughout their lives, thereby, giving a way toward blending of innovation with ecological responsibility (Kumar et al. in 2023 international conference on disruptive technologies (ICDT).

Keywords Smart technology · Sustainability · Unified payments interface (UPI) · Environmental impact · Financial inclusion

13.1 Introduction

In the era marked by escalating environmental concerns along with rapid technological advancements have taken place. The convergence of smart technology and sustainability have emerged as a critical focus for the policymakers, businesses, institutions and investors worldwide. This chapter is focused on the current practices around India's Unified Payments Interface (UPI) and global initiatives.

13.1.1 The Necessity of Sustainable Technology

Smart sustainable technologies evolved through digital revolution of the late twentieth century. Simple computerization has transformed into a sophisticated web of interconnected systems and processes efficient of autonomous decision-making. This change has profoundly affected resource handling and has greatly supported protection of environment. Internet of Things (IoT) has also emerged as one of the foundations of smart sustainable technology, with the use of billions of sensors used repeatedly for collection of data and air quality measurements, to energy consumption patterns (Kumar et al. 2023a, b, c, d, e).

One of the most promising applications of smart technology can be found in energy management. Smart grids represent a fundamental shift from traditional power distribution systems, offering real-time load balancing capabilities, efficient integration of renewable energy sources, dynamic pricing mechanisms, and predictive maintenance to reduce system failures and energy waste. In urban environments, smart cities demonstrate the broad and widespread application of sustainable technology, particularly in waste management and water conservation (Kumar et al. 2024a).

The successful implementation of smart sustainable technologies relies heavily on strong policy frameworks. Government initiatives should prioritize investments in infrastructure and funding for research and development. As we progress, the ongoing advancement of smart technologies, along with increasing environmental awareness and supportive policies, will be crucial in shaping a sustainable future (Kumar et al. 2024b).

13.1.2 *Smart Technology Investments*

Investments in smart technology cover various types of initiatives, including renewable energy systems, advanced logistical networks, precision agriculture, smart urban infrastructure etc. These investments go beyond just adopting the latest technologies; they also signify a significant change in our development approach, emphasizing long-term sustainability rather than immediate profits. The global market for such smart technology aimed at sustainability is expected to see considerable growth, with some projections indicating it could hit $166.7 billion by 2028 (BCC Research 2021).

Investing in smart technology marks a significant change in development strategy, focusing on long-term sustainability rather than short-term profits across various sectors like renewable energy, smart transportation, precision agriculture, and urban infrastructure.

Cybersecurity has become a crucial aspect of these investments, requiring a substantial commitment to security infrastructure, regular audits, and monitoring systems.

New technologies such as quantum computing, advanced materials science, and biotechnology offer fresh avenues for sustainable development. The success of these initiatives hinges on finding a balance between immediate expenses and long-term advantages, ensuring strong security protocols, investing in workforce development, and encouraging collaboration among all parties involved.

This general approach to smart technology investment helps equip organizations and governments to tackle future challenges more effectively and seize opportunities in the rapidly changing technological environment, all while advancing environmental sustainability, economic efficiency, and social progress.

13.2 The Landscape of Smart Technology for Sustainability

13.2.1 *Smart Sustainability*

Smart technology, when it comes to sustainability, means AI or the IoT, or even blockchain, in a bid to improve the environment further through minimizing the ecological footprints. These are technologies that have been known to derive and process real-time information, act autonomously, and improve processes further by making them less energy-consuming and less harmful to the environment.

This approach is consistent with the United Nations Sustainable Development Goals (SDGs), particularly the Sustainable Development Goals number 9 on (Industry, Innovation and Infrastructure) and 11 on Sustainable Cities and Communities (United Nations 2015).

It emphasizes that environmental problems cannot be solved in an abstract vacuum but must be dealt with alongside the economic and the social context. It shows that by embedding digital technologies into sustainability strategies, institutions and the state can have environmentally sustainable and at the same time economically and socially productive states. Through smart sustainability measures, it is demonstrated how technology can be harnessed to fast track the achievement of these goals.

The field-specified instances of smart sustainability are already visible in practice across various sectors. In agriculture, the irrigation system is much optimized through AI systems that use IoT sensors to cut water and improve crop yields. AI methodologies are adopted in manufacturing facilities that bring about waste and energy reduction and boost or sustain the output ratio.

The development of blockchain has transformed the idea of smart sustainability because it provides new ways of ensuring transparency and accountability with regard to a particular environmental initiative.

This correlation of the two trends: technological development and ecological awareness shows the direction to which the evolution of the mankind is headed in terms of implementing new technologies which will help to save and restore the natural environment.

13.2.2 Critical Sectors and Smart Sustainable Technologies

1. Energy Sector

Smart technologies are essential to the energy sector's shift to renewable energy sources and energy consumption optimization. For instance, the Indian government's ambitious smart meter program aims to replace 250 million conventional meters with smart meters by 2025.

Globally, investments in smart grid technologies are projected to reach $161.1 billion by 2029 from an estimated $73.8 billion in 2024, with a CAGR of 16.9% from 2024 to 2029 (Markets and Markets 2022). These investments are driven by the need for more resilient and efficient energy systems capable of accommodating the growing portion of energy sources which are renewable.

2. Transportation Sector

Smart transportation technologies are revolutionizing mobility while reducing environmental impact. Electric vehicles (EVs), coupled with smart charging infrastructure, are at the forefront of this transformation. In India, the government's FAME II (Faster Adoption and Manufacturing of Hybrid and Electric Vehicles) scheme has allocated INR 10,000 crore to promote EV adoption and create charging infrastructure (Department of Heavy Industry, Government of India 2024).

Intelligent Transportation Systems, also called ITS are investing areas as they integrate sensors, communication technologies, and data analytics to improve traffic management, and combat congestion and emissions. The global ITS market is projected to grow from \$50.7 billion in 2024 to \$70.7 billion by 2029, at a 6.9% CAGR (MarketsandMarkets 2019).

Smart transportation technologies are also giving birth to new paradigms for mobility, especially Mobility-as-a-Service (MaaS) platforms which combine disparate transport modes under a single umbrella that is easy for end-users. These systems deploy artificial intelligence to design the routing, anticipate demand, and allow for easy interchanges between transport modes aimed at further mitigating the effects of urban mobility on the environment.

The smart transportation strategies have the potential to tackle more than just environmental issues. They also help to enhance traffic safety using enhanced V2V and V2I communication systems, improve the quality of urban life by decreasing congestion and noise, and even contribute to social justice through improving reliable transportation access for every member of the society (Kumar et al. 2023c).

3. Agriculture Sector

Smart farming, a more dynamic form of agriculture, utilizes technology in its quest to sustain food production. Aging together as a collective joins this innovative transformation. One of the technological innovations allows farmers to obtain up to date information regarding nutrient uptake, temperature variation among others via IoT devices distributed throughout the farming region (Kumar and Khanna 2023).

Just how IoT does, drone technology has been doing wonders too within the realm of agriculture and this is slowly bringing together the advantages of obtaining global positioning systems all the entire area while scanning the field (fencing) using a drone. The technology not only helps in more efficient farming but also helps in the reduction of the adverse effects brought on by this farming method. The smart farming market, which was valued at USD 16.2 billion in 2023 is expected to reach around USD 25.4 billion in 2028, growing at a CAGR of 9.4%. This is due to the fact that food sustainably needs to be increased rapidly due to population increases.

Smart agricultural practices are valuable in combating food insecurity during a time when climatic change is altering farmer practices and is also vital for the future of food production as it preserves the earth's natural resources.

4. Urban Planning and Smart Cities

The mission mainly focuses on multiple key areas which includes smart water management that seeks to put in place proper systems for waste disposal as well as disposal, recycling and energy-efficient buildings aimed at cutting down use and impact on the environment. In these undertakings, technology has been effectively applied and this improves infrastructure, service provision and increases the sense of community among residents.

The Smart Cities Mission seeks in transforming urban spaces to make them not only more habitable but also more flexible and stronger to withstand future challenges.

13.2.3 Trends in Sustainable Technology Investments

The worldwide arena of sustainable technology investments is marked by rising commitments from both governmental and commercial sectors. The International Energy Agency reports that global investment in clean energy technology amounted to $1.7 trillion in 2023, markedly surpassing prior years (IEA 2023).

Corporate goals for sustainability have prompted significant investments in smart, sustainable technology. Companies like Microsoft, Google, and Amazon have committed to achieve carbon neutrality or carbon negativity by heavily investing in renewable energy, energy-efficient data centers, and advanced technologies to reduce their environmental impact (Corporate Reports 2022).

There is a growing number of firms and governments who are pledging to public reporting on emissions and climate risks, therefore allowing investors to more accurately assess the environmental effect of their portfolios. Such frameworks promote the alignment of finance with climate-positive initiatives to foster a more sustainable global economy.

Multilateral development banks are progressively financing sustainable initiatives in developing economies, emphasizing renewable energy, sustainable agriculture, and low-carbon urban development. These expenditures signify the pressing necessity for low- and middle-income nations to achieve sustainability objectives while tackling distinct regional issues.

In conclusion, the landscape of smart technology for sustainability is vast and ever evolving. Energy, transportation, agriculture, and even urban planning today offer fresh answers to many environmental challenges, owing to the advent of smart technology.

13.2.4 Indian Scenario

India's sustainable technology ecosystem is quickly expanding, thanks to government regulations, commercial initiatives, and a growing climate consciousness. Initiatives such as the National Action Plan on Climate Change (NAPCC) and the FAME plan promote sustainable energy and electric vehicle adoption, with India aiming for 50% renewable energy by 2030.

Companies such as Reliance Industries and Tata Group have made significant investments in green technology; for example, Reliance has committed $10 billion to renewable energy and green hydrogen enterprises. India's climate technology startup ecosystem is thriving. Companies such as ReNew Power are at the forefront of renewable energy, while BluSmart is a leader in sustainable materials. Investments in Indian climate tech businesses will exceed $1 billion by 2023, indicating increased investor interest (Tracxn 2023).

Initiatives such as the Smart Cities Mission as well as digital innovations like the Unified Payments Interface, have laid the groundwork for India's efforts to promote urban sustainability and digital transformation. The expansion of financial inclusion via the seamless and interoperable mode of digital transactions across bank networks has been large and remarkable through UPI, especially in areas under-owned and under-controlled in terms of economic participation. Lessening the paper-based drudgery in this whole process, the system indirectly cuts through the demerits which have been associated with environmental pollution stemming from conventional banking.

13.3 UPI-Transformative Financial Technology Adoption

13.3.1 Impact of UPI on Digital Payments

Unified Payments Interface or UPI is a real-time payment system that was created by the National Payments Corporation of India (NPCI) which enables smooth and seamless bank transactions between various banks. It was launched in 2016 and since then, UPI has absolutely transformed the digital payment scenario in India, providing an instant and cost-effective method for transferring funds.

Some of the key features of UPI:

- Instant real-time payments
- Interoperability between banks and fintech platforms
- Simple and secure transaction process with the use of virtual payment addresses
- Support for both peer-to-peer (P2P) and merchant payments.

The impact of UPI on India's digital payment ecosystem has been profound. According to NPCI data, UPI transactions have grown exponentially since its inception. During financial year 2017–18, UPI processed 92 crore transactions worth INR 8.375 crore. By financial year 2022–23, this had grown to 8.572 crore transactions worth INR 139.22 lakh crore (NPCI 2023). This remarkable growth has positioned India as a global leader in digital payments. A report by ACI Worldwide ranks India as the global leader in real-time payments, accounting for 43% of such transactions worldwide in 2021 (ACI Worldwide 2022).

13.3.2 *Digital Payments and Sustainability*

While the primary goal of UPI was to promote financial inclusion and reduce the dependence on cash transactions, it has inadvertently contributed to environmental sustainability in several ways:

- Reduction in Physical Currency Production: The shift toward digital transactions has reduced the need for physical currency production. The Reserve Bank of India (RBI) reported a decrease in currency notes in circulation from 90.27% of total payments in FY 2015–16 to 83.31% in FY 2020–21 (RBI 2022). This reduction in physical currency production translates to lower resource consumption and reduced environmental impact associated with paper and ink production.
- Lower Costs for Moving and Delivering Money: Using less cash means one doesn't need to move and hand out as much real money. This leads to using less fuel and putting out fewer greenhouse gases from trucks that carry cash around.
- Using Less Power at Cash Machines: People aren't using ATMs as much. Instead, they're doing more of their money stuff on their phones. According to Singh et al. (2016), a 10% drop in ATM transactions might save almost 32 million kWh of energy per year in India.
- Ability to Develop Green Fintech Innovations: Among the developments provided by UPI, various applications developing to promote renewable projects based on transaction round-off to invest disparities or pay for carbon emissions have received widespread approval (PwC 2024).

13.3.3 *Challenges and Issues in Sustainable Technology Investments*

13.3.3.1 Technological Barriers and Solutions

Smart tech that's good for the planet faces a bunch of tech hurdles that slow down how much people use it and how well it works. One big problem is getting different systems to work together and follow the same rules. A lot of smart setups work on their own, which makes it hard for them to talk to each other and share info well. When systems don't play nice, it leads to a mess of separate parts, wastes time and money, and makes people less likely to use them when you've got a lot of different groups and systems involved.

People are also worried about keeping data safe and private. This is because smart tech often deals with tons of sensitive info in areas like healthcare and running cities.

13.3.3.2 Economic Considerations and Return on Investment

The economic factors also are critical in the acceptance of the sustainable smart technologies since one of the major challenges that must be faced is the high initial cost associated with the sustainable technologies. Most of the times these require a huge initial investment in equipment, infrastructure, and highly trained personnel, which may be unaffordable for many areas or companies with limited budgets.

Uncertain return on investment is another challenge, since the long-term benefits of sustainable technologies–although sizeable–are not always immediately apparent. The market distortions in support of fossil fuels and unsustainable practices further complicate the economic picture in regard to sustainable technologies. These handouts artificially lower the prices of traditional energy sources and make it hard for sustainable alternatives to compete.

Addressing such economic hurdles through financing models, full-blown ROI methods, and funding changes will make sustainable smart technology usage economically viable; thus, its larger implementation will be more effective across sectors.

13.3.3.3 Social and Cultural Factors Influencing Adoption

The social and cultural variables play major factors in the adoption of sustainable smart technologies since societal awareness, behavioral change, and equal access serve as immediate determinants of the success of such innovations. Among the major barriers, one is the general lack of awareness and knowledge of the benefits and functions of sustainable technologies. For instance, the UJALA scheme in India achieved huge success in the distribution of LED bulbs, partly because it had solid awareness creation toward energy savings and cost benefits from using LEDs; hence, it saw wide adoptions. Behavioral change poses another critical challenge. For instance, smart energy devices or waste recycling habits are not very convenient at first. Smart meters in this regard show real-time feedback on energy uses, which implicitly persuades users to adopt energy-saving behaviors through visual cues and data transparency to develop more sustainable behavior over time (Carroll et al. 2014).

Since limited access to digital tools and Internet services in a lot of underserved regions puts a huge strain in the adoption of sustainable technologies, this also creates a lot of digital divide and therefore negatively impacts the adoption.

13.3.4 Opportunities in Smart Technology Investments

While the challenges in sustainable technology investments are substantial, they also create a wealth of opportunities that can drive economic, social, and environmental progress. The push to overcome technological hurdles—such as energy intermittency in renewable sources like wind power and solar power, has sparked substantial advancements in energy storage technologies.

The maturation of sustainable technologies opens new markets and encourages in developing new and innovative business models. For instance, with the rise of electric vehicles, the infrastructural support of EVs, like charging stations, battery recycling, and maintenance, started seeing manifold growth in demand. Companies that will be able to adapt to the new market conditions and start investing in sustainability solutions will enjoy serious strategic advantage and leading position in this emergent business.

Furthermore, the investment in smart technology increases the efficiency of resources, creating unparalleled opportunities to optimize resource use.

Furthermore, sustainable technologies like distributed renewable energy systems, such as roof-top solar panels and community-based wind farms, offer enhanced flexibility, enabling communities to bear climate-related disruptions and natural disasters.

Therefore, the range of opportunities taken from sustainable smart technology investments shows its potential to make economies transition in the direction of adopting a more durable and sustainable global future.

13.4 Adoption of Sustainable Smart Technology Investments

The successes, such as UPI in India and many other initiatives undertaken globally, have shown the potential of these smart technologies to contribute toward sustainability. This, however, will need a balanced investment approach that looks at innovation, scalability, and long-term sustainability. This section explores some of the key investment strategies for sustainable smart technologies from Indian and global experiences (Kumar et al. 2023a, b, c, d, e).

13.4.1 Government Policy and Promotion

Government policy and promotion are thus at the very heart of the trajectory taken by investment in sustainable technologies, providing stimuli for innovation, financing, and implementation. Therefore, in this section we will cover various initiatives and financial incentives which are required in order to support sustainable development and for that we will pay particular attention to the various national missions and programs, regulatory standards and financial measures.

13.4.2 National Missions and Programs

In order to help countries tackle environmental challenges and to promote economic growth in India, national missions and programs provide the strategic bases for it. In India, the National Smart Cities Mission and the National Solar Mission are some of the important initiatives undertaken for sustainable urban development and renewable energy. The initiative called "Smart Cities Mission" which was launched in 2015, aims to develop 100 smart cities across India, in order to emphasize the integration of smart technology to enhance the urban living conditions. By July 2024, the program had successfully channeled over INR 1,44,237 crore into various projects, reflecting the government's commitment to sustainable urban planning and infrastructure (Ministry of Housing and Urban Affairs, Government of India 2021).

At the same time, the National Solar Mission has also helped India in advancing India's renewable energy landscape. As India's installed solar capacity has now surpassed 70.1 GW in 2023, the mission has played a very important role in attracting multiple investments and facilitating the growth of Indian solar energy sector (New and Renewable Energy, Government of India 2021). As per India's commitment to Paris Agreement, this mission is a broader strategy in showcasing how the national programs can align with the global sustainability goals. In the same way, a lot of similar initiatives have been undertaken to promote sustainable technology investments (Kumar et al. 2023a, b, c, d, e). For example, The European Green Deal outlines EU's commitment in order to achieve climate neutrality by 2050, putting in at least €1 trillion in sustainable investments in the next 10 years. This initiative aims to hugely and positively change Europe's economy by investing in clean energy, sustainable agriculture etc. (European Commission 2020).

13.4.3 Regulatory Frameworks and Standards

For Investors in India regulatory frameworks and standards are very crucial components when it comes to promoting sustainable technology investments as they provide clarity for investors. Here the energy conservation building code or ECBC enables minimum energy performance standards for setting up a new commercial building, and this helps in driving up the investments in energy-efficient technologies (Bureau of Energy Efficiency, Government of India 2017). Therefore, by mandating these energy efficiency measures not only is the building performance enhanced by the ECBC but it also contributes to very high energy savings across the whole commercial sector.

By classifying activities based on their sustainability credentials European Union's Sustainable Finance taxonomy assessed the investors for recognizing projects that align with their environmental objectives (European Commission 2020).

Also, international cooperation in developing regulatory standards can further enhance the effectiveness of these frameworks. To promote best practice exchange partnerships between countries and organizations is highly required. It will also help them aligning the global regulations and reduce investment barriers that makes it easier for all involved corporations to adopt to the sustainable technologies across borders.

13.4.4 Fiscal Incentives and Subsidies

Financial incentives and subsidies also play a very important role in motivating governments to increase or boost investments aimed for sustainable technologies. A prime example of that would be the 2021 Production Linked Incentive or PLI scheme for advanced chemistry cell (ACC) battery storage that was launched by the Ministry of Heavy Industries. This initiative offers incentives to promote domestic production of various Battery storage technologies which is very important when it comes to integration of renewable energy and supporting electric mobility. On a global scale other countries too have implemented similar financial policies to encourage investments in the renewable energy sector. Germany's Renewable Energy Sources Act (EEG) Incentivizes renewable energy adoption through feed in tariffs to the energy producers (German Federal Ministry for Economic Affairs and Energy 2021). This policy has hugely driven Germany's renewable energy expansion and help them in positioning themselves as a leader in clean energy technology. Other than just the direct financial incentives alternative financial mechanisms such as green bonds and climate finance are also gaining a lot of momentum these days. By providing options of various funding types these approaches make it easy for a wide range of sustainable projects to secure financial aid, facilitating and giving chance to a more inclusive transition toward a green economy.

13.4.5 Role of Private Sector

The private sector is increasingly seen as a key driver in the shift toward sustainable technology, with businesses leading the charge by sustainability commitments, creative financing, and strategic investments in green technologies. As consumer demand for eco-friendly practices grows and regulations tighten, many companies are stepping up with significant sustainability pledges, reshaping markets and fueling large-scale investments in clean tech.

Take India, for example. Reliance Industries has pledged to invest INR 75,000 crore in clean energy over the next three years, a move that could serve as a major catalyst for sustainable development in the country (Reliance Industries 2021). This massive investment will not only enhance the company's renewable energy capabilities but also improve the overall efficiency of its energy use and help make its operations more sustainable across various sectors of its business.

On a global scale, companies like Microsoft are making even more ambitious commitments, such as becoming carbon negative by 2030. This goal is driving investments in carbon reduction and removal technologies (Microsoft 2020). The shift toward sustainability is changing the business landscape, with companies being pushed to innovate and invest in clean technologies that have a positive environmental impact.

Corporate venture capital is also emerging as a vital source of funding for sustainable technologies. Major corporations like BP and Shell have venture arms focused on investing in clean energy startups (BP Ventures 2021; Shell Ventures 2021). These corporate venture funds play a very essential role in accelerating the development of sustainable innovations, offering both capital and strategic support along with industry expertise and insights into employee well-being (Singh et al. 2021).

In conclusion, the private sector is central to the growth of sustainable technology. Through sustainability commitments, innovative financing models, and corporate venture capital, businesses are driving the transition toward a more sustainable future, opening up new economic opportunities and proving that sustainability is not only vital for the planet but also key to long-term growth and innovation.

13.4.6 Public Private Partnership Initiatives

The private sector is increasingly at the forefront of driving the shift to sustainable technology. Companies are making bold sustainability commitments, exploring new financing models, and investing strategically in clean technologies. With growing consumer demand for greener practices and tighter regulations, many businesses are stepping up, making major sustainability pledges that could reshape markets and lead to significant investments in cutting-edge solutions.

Take India, for example. Reliance Industries has announced plans to invest INR 75,000 crore in clean energy projects over the next three years. This massive investment is expected to drive sustainable development, boosting renewable energy efforts, enhancing energy efficiency, and promoting more sustainable operations across its diverse business sectors.

On the global stage, companies like Microsoft are setting ambitious goals, such as becoming carbon negative by 2030. This bold move is not just about reducing emissions but also about investing heavily in technologies to remove carbon from the atmosphere. Such commitments are reshaping industries, pushing businesses to innovate and find new ways to reduce their environmental footprint while also tapping into growth opportunities.

Corporate venture capital is also playing a significant role in funding sustainable tech. Major companies, including energy giants like BP and Shell, have established venture funds focused on clean energy solutions. These funds provide essential capital and strategic support, helping to accelerate the development of new technologies, while also leveraging industry expertise to drive innovation and improve employee welfare.

In the end, the private sector is becoming a key driver of the transition to a more sustainable future. By making sustainability a priority, exploring innovative financing models, and supporting new ventures, businesses are not only opening up new economic possibilities but also demonstrating that sustainability is vital for long-term growth and innovation.

13.4.7 Role of Communities and Individuals

The involvement of communities and individuals is crucial for adopting and succeeding with sustainable smart technologies. Local initiatives, like community-run solar cooperatives and waste management programs, help people actively contribute to sustainability. These efforts can be seen in several Indian states and other global clean energy projects (Shiradkar et al. 2023; European Commission 2022). Additionally, individual choices, such as using energy-efficient appliances or digital payment systems like UPI, can have a large environmental impact. These actions not only support sustainability but also create demand for eco-friendly products and services (National Payments Corporation of India 2024).

Therefore, empowering people through education, incentives, and supportive policies is essential for reaching sustainability goals. Learning from successful case studies and global best practices can offer useful insights for this effort.

13.5 Future Scenario

Looking toward the future of smart technology investments for a sustainable environment, it's crucial to draw on insights from global trends, real-world case studies, and valuable lessons from initiatives like India's UPI.

13.5.1 Emerging Sustainable Technologies and Their Applications

The rapid growth of Artificial Intelligence (AI) and Machine Learning (ML) offers exciting possibilities for improving the sustainability of various industries. These technologies can go through massive amounts of data in no time, thus helping us to make more informed decisions in areas such as climate forecasting, energy management, and precision agriculture. To unlock AI's full potential in advancing sustainability, funding for research should be directed at tackling the most pressing environmental challenges. At the same time, it's crucial to establish clear ethical guidelines to ensure that AI operates transparently, fairly, and responsibly.

Blockchain technology also stands out as a powerful tool for sustainability, particularly in areas like supply chain transparency, carbon credit trading, and the renewable energy sector. By securely tracking transactions, blockchain can help increase accountability and trust. Additionally, the Internet of Things (IoT) offers an effective way to monitor environmental changes in real-time. IoT devices can gather key data on air quality, water usage, and energy consumption, providing insights that drive better sustainability practices. To fully tap into this potential, efforts should be focused on creating affordable, energy-efficient IoT sensors that can be easily scaled for widespread use.

13.5.2 Imperative Strategies for Future Success

Creating flexible and adaptable solutions is crucial for implementing sustainable smart technologies in various settings. Modular components for smart cities allow for gradual adoption, letting cities integrate these technologies at their own pace and expand them as needed.

Innovative financing methods are also key to speeding up the adoption of sustainable smart technologies. Promoting green bonds and sustainability-linked bonds provides a robust funding platform for large-scale projects aimed at sustainability. These financial tools help stakeholders support environmentally friendly projects while offering good returns on investment.

Addressing the skills gap in sustainable technology is vital for the successful implementation of smart systems. This involves combining sustainability with digital skills through education, preparing future generations for innovation. Specialized training and courses will help employees in industries like energy, agriculture, and city planning to effectively master new technologies.

13.5.3 Role of International Cooperation and Sharing of Best Practices

International co-operation and knowledge exchange play a fundamental role in enhancing sustainable smart technologies. Best-practice platforms and technology transfer provide cross-border access to new innovative solutions protected by proprietary rights, therefore encouraging innovation and co-operation. Joint research between institutions in developed and developing nations can further build on this, creating partnerships that enhance capacity in technology and solve localized problems.

International cooperation can take the form of global promotion of standards and interoperability in sustainable smart technologies, among other areas. There is a need for interoperability to ensure that systems and platforms can integrate fully as this maximizes the full benefits of smart technologies. The adoption of open standards in government purchases will make room for sustainable technology adoption in the reduction of entry barriers. Global sustainability benchmarks for countries will help in appraising progress and sharing the best practices to encourage continual improvement.

International cooperation is necessary to deal with the global challenges. With the help of cooperation, much can be added to efforts. Developing global platforms that coordinate climate change responses with the help of smart technologies will enable nations to cooperate better. These hubs will become centers for information, strategy, and resource sharing. International cooperation and sharing of best practices are required for better implementation of sustainable smart technologies so that all nations can contribute to and reap from a greener future.

13.6 Conclusion

Smart tools and a sustainable future: Development and application Smart technologies weave in the way of a healthy future. Diverse examples have been followed throughout this chapter-from India's Unified Payments Interface to global efforts in green energy, smart cities, and sustainable agriculture-all of which show how powerful smart technologies can be. These innovations are not only system improvements but a mindset shift that can drive big positive change across sectors, which can improve efficiency, use resources better, and also be more inclusive.

Achieving this potential depends on a broad approach that includes strategic investments, supportive policies, and a consistent effort to address challenges like interoperability, data security, and rightful access.

While being innovative is the key to success, solutions have to scale and adapt to various settings. To this end, there are a few significant things that have to happen. First, a vision must be underpinned with sustainable policies and regulations that should create stability and predictability for investors and stakeholders in the long run. Strong partnerships must be set between governments, private sector organizations, academic institutions, and local communities to marshal all kinds of expertise and resources. It should involve the development of local capacity and skills such that over time, these communities could effectively implement and sustain such initiatives.

In addition, there is a need for open standards and interoperability that will ensure maximum impact and reach of sustainable technologies while facilitating collaboration and integration across platforms. Innovative financing mechanisms are also important as they will support large-scale projects and grassroots initiatives that can allow diverse stakeholders to access the resources necessary. International cooperation will be a very important part of the fight against global issues, spreading knowledge, and multiplying successful practices across the globe. Thus, investments in smart technology serve as significant tools for building a more sustainable, egalitarian, and prosperous future.

Funding NA.

Conflict of Interest: There is no conflict of interest.

References

Benefits of oneM2M (n.d.) oneM2M. https://www.onem2m.org/using-onem2m/what-is-onem2m/

Bhushan B (2024) India's smart meters revolution: Dr. Bharat Bhushan on growth and innovation. TimesTech. https://timestech.in/indias-smart-meters-revolution-dr-bharat-bhushan-on-growth-and-innovation/

bp ventures | What we do | bp America (n.d.) bp America. https://www.bp.com/en_us/united-states/home/what-we-do/innovation-and-engineering/bp-ventures.html

Budinski M (2023) A green cloud race: Microsoft, Google, and Amazon compete for a sustainable future. Deployflow. https://deployflow.co/blog/a-green-cloud-race-microsoft-google-and-amazon-compete-for-a-sustainable-future/

Carroll J, Lyons S, Denny E (2014) Reducing household electricity demand through smart metering: the role of improved information about energy saving. Energy Econ 45:234–243. https://doi.org/10.1016/j.eneco.2014.07.007

Climate Bonds Initiative (2022) $500bn green issuance in 2021: social and sustainable acceleration; annual green $1tn in sight for market. https://www.climatebonds.net/2022/01/500bn-green-issuance-2021-social-and-sustainable-acceleration-annual-green-1tn-sight-market

Delhi Metro plans to get 50 per cent of its energy from solar power by 2031 (2023) Times Now. https://www.timesnownews.com/delhi/delhi-metro-plans-to-get-50-per-cent-of-its-energy-from-solar-power-by-2031-article-99323607

Denmark sources record 47% of power from wind in 2019 (2019) Reuters. https://www.reuters.com/article/business/environment/denmark-sources-record-47-of-power-from-wind-in-2019-idUSKBN1Z10KB/

Dey A (2023) No funding slowdown for climate tech startups; companies bag $1 billion funding in first half of 2023. The Times of India. https://timesofindia.indiatimes.com/business/india-business/no-funding-slowdown-for-climate-tech-startups-companies-bag-1-billion-funding-in-first-half-of-2023/articleshow/101464302.cms

Digital finance: A gateway to environmental sustainability (n.d.) IJU Journal. http://journal.iujharkhand.edu.in/June-2023/Digital-Finance-A-Gateway-to-Environmental-Sustainability.html

Dumoulin A (2023) A brief summary of the World Bank's Carbon Pricing Report 2023. https://www.ciat.org/a-brief-summary-of-the-world-banks-carbon-pricing-report-2023/?lang=en

Energiewende: Germany's planned transition to a low-carbon, nuclear-free economy—Pathfinders (2023) Pathfinders. https://www.sdg16.plus/policies/energiewende-germanys-planned-transition-to-a-low-carbon-nuclear-free-economy/

Energy Conservation Building Code (ECBC) | Bureau of Energy Efficiency, Government of India, Ministry of Power (n.d.). https://beeindia.gov.in/en/energy-conservation-building-code-ecbc

EU taxonomy for sustainable activities (n.d.) Finance. https://finance.ec.europa.eu/sustainable-finance/tools-and-standards/eu-taxonomy-sustainable-activities_en

Finance and the Green Deal (n.d.) European Commission. https://commission.europa.eu/strategy-and-policy/priorities-2019-2024/european-green-deal/finance-and-green-deal_en

Global Climate Tech Venture Capital Report—Full Year 2021 (n.d.). https://www.holoniq.com/notes/global-climatetech-vc-report-full-year-2021

Gupta A (2024) India's path to decarbonization: a strategic roadmap to achieve net zero by 2030. CXO Today. https://cxotoday.com/story/indias-path-to-decarbonization-a-strategic-roadmap-to-achieve-net-zero-by-2030/

In focus: Energy communities to transform the EU's energy system (2022) Energy. https://energy.ec.europa.eu/news/focus-energy-communities-transform-eus-energy-system-2022-12-13_en

India surges ahead as the world's leader in real-time payments—Boosting economic growth—ACI Worldwide report|ACI Worldwide (n.d.) ACI Worldwide. https://investor.aciworldwide.com/news-releases/news-release-details/india-surges-ahead-worlds-leader-real-time-payments-boosting

Intelligent Transportation System Market Size, Share & Trends [2024] (2024) MarketsandMarkets. https://www.marketsandmarkets.com/Market-Reports/intelligent-transport-systems-its-market-764.html

Johny RM (2023) At 8.7 billion, UPI hits record transactions in March. Hindustan Times. https://www.hindustantimes.com/business/at-8-7-billion-upi-hits-record-transactions-in-march-101680411218117.html

Kumar R, Khanna R (2023) RPA (robotic process automation) in finance & accounting and future scope. In: 2023 second international conference on augmented intelligence and sustainable systems (ICAISS). IEEE, pp 1640–1645

Kumar R, Sexena A, Gehlot A (2023a) Artificial intelligence in smart education and futuristic challenges. In: 2023 international conference on disruptive technologies (ICDT). IEEE, pp 432–435

Kumar R, Lande A, Kumar D, Malhotra RK, Sharma A (2023b) Technology bridging-in entrepreneurs and consumers in product development. In: 2023 11th international conference on intelligent systems and embedded design (ISED). IEEE, pp 1–5

Kumar R, Saxena A, Singh R (2023c) Robotic process automation bridge-in banking institute and consumers. In: 2023 international conference on disruptive technologies (ICDT). IEEE, pp 428–431

Kumar R, Kandpal B, Ahmad V (2023d) Industrial IoT (IIOT): security threats and countermeasures. In: 2023 international conference on innovative data communication technologies and application (ICIDCA). IEEE, 829–833

Kumar R, Malholtra RK, Grover CN (2023e) Review on artificial intelligence role in implementation of goods and services tax (GST) and future scope. In: 2023 international conference on artificial intelligence and smart communication (AISC). IEEE, pp 348–351

Kumar V, Kumar R, Goel R, Singh T, Kaur MJ (2024a) The evolution of digital marketing technologies: a fuzzy TOPSIS approach to selected technology in E-commerce platform. Int J Syst Assur Eng Manag 1–14

Kumar R, Goel R, Singh T, Mohanty SM, Gupta D, Alkhayyat A, Khanna R (2024b) Sustainable finance factors in Indian economy: analysis on policy of climate change and energy sector. Fluctuation Noise Lett 23(2)

Lighting a billion: The UJALA program's transformational impact in India—CLASP (2024) CLASP. https://www.clasp.ngo/report/net-zero-heroes/case-studies/lighting-a-billion-the-ujala-programs-transformational-impact-in-india/

Massachusetts Institute of Technology (2018) The convergence of digitalization and sustainability|MIT Sloan Management Review. MIT Sloan Management Review. https://sloanreview.mit.edu/article/the-convergence-of-digitalization-and-sustainability/

Mohanty SN, Singh T, Goel R, Baral SK, Kumar R (2024) A study on building awareness in cyber security for educational system in India using interpretive structural modellings. Int J Syst Assur Eng Manag 1–11

Pandit P (2024) Digital India: bridging the divide in the world's largest democracy. Vision of humanity. https://www.visionofhumanity.org/digital-india-bridging-the-divide-in-the-worlds-largest-democracy

PLI scheme for national programme on advanced chemistry cell (ACC) battery storage|Ministry of Heavy Industries (n.d.) Ministry of heavy industries. https://heavyindustries.gov.in/pli-scheme-national-programme-advanced-chemistry-cell-acc-battery-storage

Reliance Industries Limited—Retail markets|Telecom|Petroleum refining & marketing|Petrochemicals|Hydrocarbon exploration & production|Jio 4G|Reliance shares (n.d.) Reliance Industries Limited. https://www.ril.com/businesses/new-energy-materials
Reserve Bank of India—Annual report (2022) Reserve Bank of India. https://rbi.org.in/scripts/AnnualReportPublications.aspx?Id=1350
Routray BB (2024) Leveraging technology to address challenges in Kenya's coffee farming sector. CSM Tech. https://csm.tech/leveraging-technology-to-address-challenges-in-kenyas-coffee-farming-sector
Shell ventures|Shell global (n.d.) Shell global. https://www.shell.com/what-we-do/technology-and-innovation/innovate-with-shell/shell-ventures.html
Shiradkar S, Sharma R, Choudhary D, Venkateswaran J, Kumar P, Solanki CS (2023) Can community based solar energy initiatives deliver on women's empowerment in India? Evidence from rural Assam, Bihar, Jharkhand and Uttar Pradesh. Energy Res Soc Sci 104:103225. https://doi.org/10.1016/j.erss.2023
Singh G, Dhanny HS, Garg V, Sharma S (2021) HR acumen in industry 4.0: managing talent and achieving balance in life. In: Financial intelligence in human resources management. Apple Academic Press, pp 1–20
SKILL COUNCIL for GREEN JOBS—Skill, Solar, Suryamitra, Green Jobs, Skill India, PMKVY, NSDC, Renewable, Capacity Building, Swachh Bharat, National Solar Mission, Wind Energy, Waste Management, Training, Assessment, QP, Qualification Packs, Sustainable, Environment, SDG's (n.d.) Skill council for green jobs. https://sscgj.in/
Smart agriculture market size, share, industry technology and growth analysis 2032 (2024) MarketsandMarkets. https://www.marketsandmarkets.com/Market-Reports/smart-agriculture-market239736790.html
Smart cities mission extended till March 2025 (n.d.) Press Information Bureau. https://pib.gov.in/PressReleaseIframePage.aspx?PRID=2030491
Smart grid market size, share and industry analysis [2024–2030] (n.d.) MarketsandMarkets. https://www.marketsandmarkets.com/MarketReports/smart-grid-market208777577.html
Solar energy corporation of India limited(SECI), a government of India enterprise under Ministry of New and Renewable Energy (n.d.) Solar Energy Corporation of India Limited (SECI). https://www.seci.co.in/web-data/docs/final-VGF_750MW_Guidelines_for-grid-solar-power-projects
Solar overview | Ministry of New and Renewable Energy|India (n.d.) Ministry of New and Renewable Energy. https://mnre.gov.in/solar-overview/
Sood D, Belgavi V, Gandhi M, Goswami S (2024) FinTech—Powering India's USD 5 trillion economy. In: PwC. https://www.pwc.in/assets/pdfs/industries/powering-indias-usd-5-trillion-economy-by-fostering-innovations.pdf

MIX
Papier aus verantwortungsvollen Quellen
Paper from responsible sources
FSC® C105338

If you have any concerns about our products,
you can contact us on
ProductSafety@springernature.com

In case Publisher is established outside the EU,
the EU authorized representative is:
Springer Nature Customer Service Center GmbH
Europaplatz 3, 69115 Heidelberg, Germany

Printed by Libri Plureos GmbH
in Hamburg, Germany